MEMOIRS
of the
American Mathematical Society

Number 459

Contractive Projections in C_p

Jonathan Arazy
Yaakov Friedman

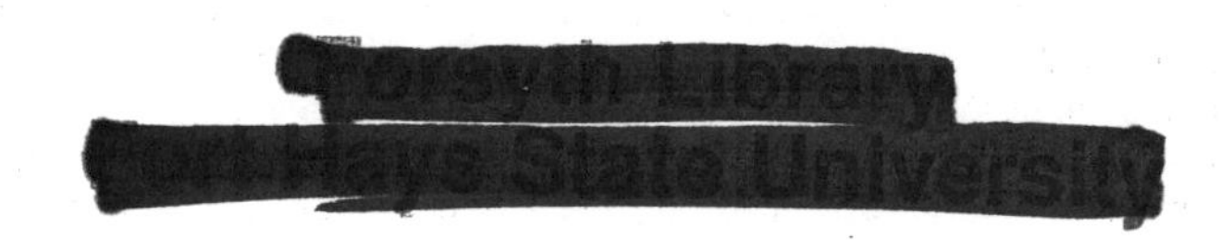

January 1992 • Volume 95 • Number 459 (first of 4 numbers) • ISSN 0065-9266

American Mathematical Society
Providence, Rhode Island

1991 *Mathematics Subject Classification.*
Primary 47D25, 47D15, 46B20, 47B10.

Library of Congress Cataloging-in-Publication Data

Arazy, Jonathan, 1942–
Contractive projections in Cp/Jonathan Arazy, Yaakov Friedman.
p. cm. – (Memoirs of the American Mathematical Society, ISSN 0065-9266; no. 459)
Includes bibliographical references (p.).
ISBN 0-8218-2515-1
1. Linear operators. 2. Hilbert space. I. Friedman, Yaakov, 1948– . II. Title. III. Series.
QA3.A57 no. 459
[QA329.2]
510 s–dc20 91-36296
[515$'$.7246] CIP

Subscriptions and orders for publications of the American Mathematical Society should be addressed to American Mathematical Society, Box 1571, Annex Station, Providence, RI 02901-1571. *All orders must be accompanied by payment.* Other correspondence should be addressed to Box 6248, Providence, RI 02940-6248.

SUBSCRIPTION INFORMATION. The 1992 subscription begins with Number 459 and consists of six mailings, each containing one or more numbers. Subscription prices for 1992 are $292 list, $234 institutional member. A late charge of 10% of the subscription price will be imposed on orders received from nonmembers after January 1 of the subscription year. Subscribers outside the United States and India must pay a postage surcharge of $25; subscribers in India must pay a postage surcharge of $43. Expedited delivery to destinations in North America $30; elsewhere $82. Each number may be ordered separately; *please specify number* when ordering an individual number. For prices and titles of recently released numbers, see the New Publications sections of the NOTICES of the American Mathematical Society.

BACK NUMBER INFORMATION. For back issues see the AMS Catalogue of Publications.

MEMOIRS of the American Mathematical Society (ISSN 0065-9266) is published bimonthly (each volume consisting usually of more than one number) by the American Mathematical Society at 201 Charles Street, Providence, Rhode Island 02904-2213. Second Class postage paid at Providence, Rhode Island 02940-6248. Postmaster: Send address changes to Memoirs of the American Mathematical Society, American Mathematical Society, Box 6248, Providence, RI 02940-6248.

Printed in the United States of America.
The paper used in this book is acid-free and falls within the guidelines established to ensure permanence and durability. ♾

10 9 8 7 6 5 4 3 2 1 97 96 95 94 93 92

Contents:

Abstract

For $1 \leq p < \infty$ let C_p be the von Neumann-Schatten class, i.e. the Banach space of all compact operators x on a separable infinite dimensional complex Hilbert space H, for which

$$\|x\|_p := (\text{trace } (x^*x)^{p/2})^{1/p}$$

is finite. In this work we establish the following result.

MAIN THEOREM Let X be a closed subspace of $C_p, 1 < p < \infty, p \neq 2$. Then the following four properties are equivalent:

(1) X is the range of a contractive projection from C_p;

(2) X is the ℓ_p-sum of subspaces, each of which is canonically isometric to the C_p ideal of a Cartan factor of one of the types I-IV;

(3) $X^{p-1} := \{v(x)|x|^{p-1}; x \in X\}$ is a closed linear subspace of C_q, $(p^{-1} + q^{-1} = 1)$ where $x = v(x)|x|$ is the polar decomposition of the operator x;

(4) $V := \overline{\text{span}}^{w^*}\{v(x); x \in X\}$ is closed under the triple product

$$\{u, v, w\} = (uv^*w + wv^*u)/2,$$

and is an atomic JCW^*-subtriple of $B(H)$. Moreover, X is a module over V, namely $\{VVX\} \subseteq X$ and $\{VXV\} \subseteq X$.

As a corollary we obtain that the category of C_p ideals of atomic JCW^*-triples is stable under contractive projections.

Key Words: Contractive projections, C_p, JC^*-triples, Cartan factors, support functional map, Schur multipliers, tripotents, partial isometries, conditional expectations, Peirce projections, triple functional calculus, monotone Schauder decomposition.

Acknowledgement: J. Arazy was supported in part by NSF Grant No. DMS 8702371. Y. Friedman was supported in part by BSF Grant No. 8800066 and NSF Grant No. DMS 8805256. Both authors benefited from the hospitality of the Mathematics Department of the University of California at Irvine during several visits. We are particularly grateful to Professor B. Russo for arranging these visits and for helping us in various ways. We would like to thank also Professors T. Barton, A. Lazar, B. Solel and H. Upmeier for various comments and corrections.

0. Introduction

Let H be a separable, infinite dimensional complex Hilbert space and let $B(H)$ denote the space of all bounded linear operators on H. Let $C_\infty = C_\infty(H)$ denote the subspace of $B(H)$ consisting of all compact operators.

For $1 \leq p < \infty$ let $C_p = C_p(H)$ be the Banach space of all $x \in C_\infty$ for which

$$\|x\|_p := (tr|x|^p)^{1/p}$$

is finite, where $|x| = (x^*x)^{1/2}$ and "tr" denotes the usual trace.

The spaces C_p are called "von Neumann-Schatten p-classes"; C_2 is the Hilbert space of Hilbert-Schmidt operators and C_1 is the trace class. The standard references for the basic properties of C_p are $[GK1]$, $[Si]$, $[M]$ and $[DS2]$.

This work is devoted to the study of contractive projections (i.e. norm-one idempotent operators) on C_p. Our main results are summarized in the above abstract and have already been announced in $[AF3]$. It extends and continues our work on contractive projections in C_1 and C_∞ $[AF2]$. While the results of the present work are similar to those of $[AF2]$ the methods are very different. Some of the results of the present work in the special case of finite dimensional C_p spaces, i.e. in $C_p^n = C_p(\ell_2^n)$ were announced in $[AF2]$.

This work can be viewed also as a generalization to the non-commutative setup of the known theory of contractive projections in $L_p(\mu)$-spaces (see $[L]$ and the references therein). According to this theory the following conditions are equivalent for a closed subspace X of $L_p(\mu) = L_p(\Omega, \mu), 1 < p < \infty, p \neq 2$:

(1) X is the range of a contractive projection from $L_p(\mu)$;

(2) X is a sublattice of $L_p(\mu)$;

(3) X is the isometric to some $L_p(\nu)$-space.

The contractive projection in question is known to be similar to a conditional expec-

Received by the editors January 18, 1991.

tation on $L_p(\nu)$, where (in the separable case) $\nu = |\phi|^p\mu$ restricted to the σ-algebra generated by the members of X, where ϕ is a function in X with a maximal support and the similarity operator is the operator of multiplication by ϕ (considered as an operator on $L_p(\mathrm{supp}(\phi), \mu)$).

In our setup, the non-commutativity of the multiplication is reflected in the richer and more complicated geometrical structure. Although some of the main tools and ideas of proofs are the same as in the commutative case, there are many conceptual and technical difficulties to overcome. For instance, in the non-commutative setup the process of changing the density (replacing μ by $\nu = |\phi|^p\mu$) is not possible, to mention just one of the differences.

Our Main Theorem is the non-commutative analog of the theory of contractive projections on $L_p(\mu)$ spaces. The space V becomes $\overline{span}^{w^*}\{v(x); x \in X\}$ (closure in the w^* topology of $B(H)$), and it is closed under the triple product $\{x, y, z\} := (xy^*z + zy * x)/2$, i.e. it is a JCW^*-subtriple of $B(H)$, (see section 2 for notions concerning JC^*-triples). V is a direct sum of classical Cartan factors (i.e. Cartan factors of types I-IV), and the natural contractive projection from $B(H)$ onto V is the substitute for the conditional expectation in the commutative setup. Also, the equivalence of conditions (4) and (1) in the Main Theorem is the non-commutative analog of the known fact that if X is the range of a contractive projection in $L_p(\mu)$ $(1 \le p < \infty, p \ne 2)$ then the supports of function in X form a σ-ring (or a σ-algebra if μ is σ-finite) and $\overline{\mathrm{span}}\{\mathrm{sgn}(f),\ f \in X\}$ is a sublattice of $L^\infty(\mu)$ which is closed under the triple product $\{f, g, h\} = f\bar{g}h$.

C_2 is a Hilbert space and so every closed subspace is the range of a contractive projection. Thus we focus here on C_p with $1 < p < \infty,\ p \ne 2$. In these cases the basic phenomenon is that geometrical conditions (like equality in norm inequalities and in particular membership in the range of a contractive projection) impose strong algebraic conditions on the elements involved. This is typical to both $L_p(\mu)$ and C_p, and was already used in the study of isometries and contractive projections in

$L_p(\mu)$-spaces (see $[L]$), and in the study of the geometry and isometries of C_p (see $[M], [A1], [AF1]$ and $[Y]$).

This work is related also to the theory of contractive projections on JB^*-triples (see $[FR1], [K2], [ES]$ and $[So]$). The Jordan triple product is the algebraic structure which respects the geometry (in JB^*-triples as well as in C_p). Therefore we use the JB^*-formalism in the present work. In this formalism our results have a relatively simple form, see the Main Theorem in the abstract, contrary to our work $[AF2]$ in which the ranges of contractive projections in C_1 and C_∞ are described without this formalism in a complicated way.

The structure of the work is as follows. In Section 1 we prove some preliminary results on contractive projections in C_p which depend only on strict convexity, smoothness and reflexivity (and thus hold in other Banach spaces having these properties). In particular, we characterize the ranges X of contractive projections in C_p, $(1 < p < \infty, p \neq 2)$ by the property (called "N-convexity") that the image of X under the support-functional map is a closed linear subspace of $C_q = C_p^*$ $(\frac{1}{p} + \frac{1}{q} = 1)$. This establishes the equivalence (1) $\Leftrightarrow$ (3) in the Main Theorem. Section 2 provides the background material on JB^*- and JC^*-triples which is relevant to the present work. Theorem 2.4, expressed in the JC^*-formalism, contains the essential part of the Main Theorem, namely the implication (1) $\Rightarrow$ (2). This theorem is proved in the rest of the work.

The main tools in studying the contractive projections in C_1 and C_∞ in [AF2] are the fact that contractive projections on these spaces respect the rich facial structures of their unit balls, and the intimate connection of these structures to the support projections, called Peirce projections, see [FR2]. In our present context C_p is uniformly convex and uniformly smooth (see [M], [T1]), and thus each non-trivial face of the unit ball is a single point. Instead, the basic technique is the study of the differentiation of the support-functional map $N_p : C_p \to C_q$ (for $2 < p$). This map is defined by $N_p(0) = 0$ and

$$N_p(x) = x^{p-1}/\|x\|_p^{p-2} \quad \text{if } x \neq 0,$$

where for x with the polar decomposition $x = v(x)|x|$ the powers x^α are defined by $x^\alpha := v(x)|x|^\alpha$.

In Section 3 we study first operator- and triple- differentiable functions, and express their derivatives as certain Schur multipliers corresponding to the divided differences and divided sums of the functions. We strengthen some of the known results, and study a new class of bounded Schur multipliers via the theory of integration with respect to a spectral family of projections. These differentiation results are applied to the map $N_p : C_p \to C_q$. As a corollary we get that if P is a contractive projection in C_p $(2 < p)$ with range $R(P)$ then all the operators $N_p'(x)y := d/dt\, N_p(x+ty)_{|t=0}$, corresponding to $x \in R(P)$, map $R(P)$ continuously into $R(P^*)$. Since these operators are identified as the Schur multipliers corresponding to the divided differences and sums of $f(\lambda) = sgn(\lambda)|\lambda|^{p-1}$, we obtain important information on the structure of P and P^*.

In the following Sections 4-7 we establish Theorem 2.4, thus proving the essential part of the Main Theorem. Sections 4 and 5 are devoted to the *analysis* and Sections 6 and 7 to the *synthesis*.

In Section 4 we show that a contractive projection P in C_p $(1 < p < \infty, p \neq 2)$ commutes with the generalized support projections, called Peirce projections, of elements in its range. A weaker version of this fact holds also in the context of C_1 and C_∞, and played a major role in the proofs in [AF2]. Also, P commutes with the local conjugation associated with any $x \in R(P)$. Thus P is *locally self-adjoint*. These results are formulated in Theorem 4.1. This theorem is established by constructing two Hilbertian seminorms on C_p (depending on a point $x \in R(P)$) in which P is bounded and self-adjoint. The bounded operators from C_p into C_q implementing these Hilbertian seminorms are linear combinations of the Schur multipliers $N_p'(x)y$ and $\frac{1}{i}N_p'(x)(iy)$. The operator M which intertwine these Hilbertian seminorms is a bounded Schur multiplier from C_p into itself which commute with P. This provides a very important information on the fine structure of P, and in particular establishes Theorem 4.1.

In Section 5 we establish the existence of "atoms" in $R(P)$. These are the normalized elements x in $R(P)$ so that every $y \in R(P)$ whose support is smaller than that of x is a scalar multiple of x. The atoms in $R(P)$ are the analoges of the matrix units $e_{i,j}$ in C_p, and are used in the synthesis part (Sections 6 and 7) to construct $R(P)$. We say that $x \in R(P)$ is *indecomposable* if it cannot be written as the sum of two non-zero, orthogonal elements from $R(P)$. It is relatively easy to established the existence of indecomposable elements. The existence of atoms is obtained by showing that being an atom is the same as being indecomposable. The proof involves an intermediate property of an element $x \in R(P)$ (*to be an atom in its diagonal*), the fine structure of the projection P established in Sections 3 and 4, and the formula for the second derivative of the map $x \mapsto x^\alpha$, $\alpha > 3$.

In Section 6 we analyse the various relations between atoms in ranges of contractive projections on C_p $(1 < p < \infty, p \neq 2)$. In analogy with [DF], [AF2] and [N] we introduce these basic relations between atoms. Elements x and y are *orthogonal* if they have orthogonal ranges and co-kernels, namely $x = e_{1,1} \otimes a$, $y = e_{2,2} \otimes b$ for some $a, b \in C_p$. Elements x and y are *colinear* if they can be written as $x = e_{1,2} \otimes a + e_{2,1} \otimes b$, $y = e_{1,3} \otimes a + e_{3,1} \otimes b$ for some orthogonal elements $a, b \in C_p$ with $\|a\|_p^p + \|b\|_p^p = 1$. Also y *governs* x if one can write $x = e_{1,1} \otimes a$ and $y = 2^{-1/p}(e_{1,2} \otimes a + e_{2,1} \otimes a)$ where $a \in C_p$ and $\|a\|_p = 1$. Then we show that any two "compatible", non-orthogonal atoms are either colinear or one governs the other (the "two case lemma"). We use this result to build in $R(P)$ C_p-ideals of classical Cartan factors of low dimension.

Section 7 is devoted to construction of $R(P)$ from the atoms. As $R(P)$ is the direct sum of orthogonal, indecomposable subspaces, each of which is the range of a contractive projection from C_p, there is no loss of generality in assuming that $R(P)$

is in fact indecomposable. We start with an atom $x \in R(P)$ and, as in Section 6, construct the colinear-ortho-governing (cog, for short) grid of the C_p ideal of the low-dimensional Cartan factor inside $R(P)$ which contain x. We continue the construction, and enlarge the cog-grid to a basis of $R(P)$. This shows that $R(P)$ is the C_p-ideal of a Cartan factor. The various cases stemming from the analysis in Section 6 lead to the various Cartan factors in Section 7. This completes the proof of Theorem 2.4, and thus completes the most important implication (1) $\Rightarrow$ (2) in the Main Theorem.

Section 8 is devoted to the completion of the proof of the Main Theorem, and to the derivation of some corollaries. At this stage we know that (1), (2) and (3) of the Main Theorem are equivalent. The implication (2) $\Rightarrow$ (4) is immediate, and (4) $\Rightarrow$ (2) is proved by using the techniques of [DF] and the results of Section 7. By analyzing the proofs in the previous sections we obtain the full description of the irreducible representation of the C_p-ideals of classical Cartan factors in C_p ($1 \leq p < \infty, p \neq 2$). An easy corollary of the Main Theorem is that the category of C_p-ideals of atomic JCW^*-triples is closed under contractive projections. This is the C_p-analogue of the fact that the category of JC^*-triples is closed under contractive projections ([FR1], see also [K2], [ES]). We consider the category of JCW^*-modules, and show that the canonical isometries from the C_p-ideals of the classical Cartan factors into the ranges of contractive projections in C_p are in fact morphisms of the category of modules. Finally we show that the contractive projections on C_p commute with certain multiplication operators defined via the triple product, and therefore can be viewed as conditional expectations.

Section 9 contains various corollaries on families of contractive projections. We show first that the *bicontractive projections* P on C_p ($1 < p < \infty, p \neq 2$) are the obvious ones. Then we show that among the C_p-ideals of Cartan factors the only ones with a monotone basis are those of types $I_{1,n}$ or IV_n for some n. Moreover, if the Cartan factor is of rank r, then in any monotone Schauder decomposition of the corresponding C_p-ideal there are gaps whos dimensions are at least $r-1$. We close Section 9 with some remarks and open problems.

The assumption that the underlying Hilbert space H is separable is technical and was made for convenience. Our results can be extended to the non separable case in the obvious way.

1. Properties of contractive projections on C_p which depend on smoothness, strict convexity and reflexivity.

This section is devoted to the study of certain properties of contractive projections on C_p $(1 < p < \infty)$ depending only on smoothness, strict convexity and reflexivity. The statements and the proofs extend to more general Banach spaces having these properties.

Recall first that with respect to the pairing

$$< x, y >= tr(xy^*); \quad x \in C_p,\ y \in C_q \quad (p^{-1} + q^{-1} = 1), \tag{1.1}$$

$C_p^* = C_q$. Thus C_p is reflexive. Recall that the *support functional map* $N_p : C_p \to C_q$ was defined by $N_p(0) = 0$ and

$$N_p(x) = x^{p-1}/\|x\|_p^{p-2}, \quad x \in C_p\backslash\{0\}. \tag{1.2}$$

It is known that $N_p(x)$ is the unique element of C_q for which

$$\|N_p(x)\|_q = \|x\|_p \quad \text{and} \quad < x, N_q(x) >= \|x\|_p^2. \tag{1.3}$$

In particular, C_p is smooth (i.e. the supporting hyperplane to the unit ball at each point on the unit sphere is unique), and by reflexivity also strictly convex. We refer to [Da] and [Di] for smoothness and convexity properies of norms in Banach spaces. In fact, C_p enjoys a much stronger property: it is *uniformly comvex* and *uniformly smooth*, see [T1] and [M]. The *norm of* C_p *is Gateaux differentiable* at any $0 \neq x \in C_p$, and its derivative is (up to normalization) $N_p(x)$, i.e.

$$\frac{d}{dt}\|x + ty\|_{p|_{t=0}} = Re < y, N_p(x)/\|N_p(x)\|_q >; \quad y \in C_p. \tag{1.4}$$

Also, clearly the map N_p is one-to-one and its inverse is N_q , i.e.

$$N_q(N_p(x)) = x; \quad x \in C_p. \tag{1.5}$$

Proposition 1.1 *Let $1 < p < \infty, q = p/(p-1)$ and let P be a contractive projection in C_p. Then:*

(i) $P^*$is a contractive projection in C_q;

(ii) For $x \in C_p$, $x = Px$ if and only if $N_p(x) = P^* N_p(x)$;

(iii) For $x \in C_p, x = Px$ if and only if $\|x\|_p = \|Px\|_p$.

Proof: Statement (i) is obvious. If $x = Px$ and $\|x\|_p = 1$, then $\|N_p(x)\|_q = 1$ and

$$1 = \|x\|_p^2 =< x, N_p(x) >=< Px, N_p(x) >=< x, P^* N_p(x) > .$$

By (i), $\|P^* N_p(x)\|_q \leq 1$. So the uniqueness of the support functional of x implies that $P^* N_p(x) = N_p(x)$. The other part of statement (ii) follows from this by the reflexivity of C_p and (1.5).

To prove (iii), let $x \in C_p$ be so that $\|x\|_p = \|Px\|_p = 1$. Then

$$1 = \|Px\|_p = \|P(x + Px)/2\|_p \leq \|x + Px\|_p \leq (\|x\|_p + \|Px\|_p)/2 = 1.$$

Thus, $\|x + Px\|_p = 2$. By the strict convexity of C_p we get $x = Px$. □

The *annihilator* of a subset Y of C_p is defined as usual by

$$Y^{\perp} = \{x \in C_q; < y, x >= 0 \quad \text{for every } y \in Y\}. \tag{1.6}$$

Notice that $Y^{\perp}$ is a closed linear subspace of C_q, and that by reflexivity, $Y = Y^{\perp\perp}$ if and only if Y is a closed linear subspace of C_p.

Proposition 1.2 *Let $1 < p < \infty, q = p/(p-1)$ and let Y be a closed subspace of C_p. Then Y is the range of a contractive projection from C_p if and only if $N_p(Y)$ is a closed linear subspace of C_q. Moreover, in this case ,$\ker(P) = N_p(Y)^{\perp}$ and P is the unique contractive projection whose range on C_p is Y.*

Proof: If Y is the range of a contractive projection P on C_p, the by Proposition 1.1, $N_p(Y) = P^*(C_q)$. Thus $N_p(Y)$ is a closed linear subspace of C_q.

Conversely, assume that $N_p(Y)$ is a closed linear subspace of C_q. Thus $N_p(Y) = N_p(Y)^{\perp\perp}$. If $y \in Y \cap N_p(Y)^{\perp}$ then $\|y\|_p^2 =< y, N_p(y) >= 0$. So $Y \cap N_p(Y)^{\perp} = \{0\}$. Similarly $Y^{\perp} \cap N_p(Y) = \{0\}$. Next, for $0 \neq y \in Y, z \in N_p(Y)^{\perp}$ we have

$$\|y + z\|_p \geq | < y + z, N_p(y)/\|y\|_p > | = \|y\|_p.$$

It follows that $W := Y \oplus N_p(Y)^{\perp}$ is closed, and that the projection $P : W \to Y$ defined by

$$P(y + z) = y; \; y \in Y, z \in N_p(Y)^{\perp}$$

is contractive and $\ker(P) = N_p(Y)^{\perp}$. W is all of C_p, indeed

$$W^{\perp} = (Y + N_p(Y)^{\perp})^{\perp} = Y^{\perp} \cap N_p(Y)^{\perp\perp} = Y^{\perp} \cap N_p(Y) = \{0\}.$$

Thus Y is the range of the contractive projection P on C_p with kernel $N_p(Y)^{\perp}$. To show the uniqueness of P, assume that Q is another contractive projection on C_p with $Q(C_p) = Y$. Obviously $PQ = Q$. Proposition 1.1 implies that $P^*(C_q) = Q^*(C_q) = N_p(Y)$, and so $Q^*P^* = P^*$. By reflexivity we get $P = P^{**} = (Q^*P^*)^* = P^{**}Q^{**} = PQ = Q$, as desired. □

Motivated by Proposition 1.2 we say that a closed subspace Y of C_p is **N-convex** if

$$N_q(N_p(x) + N_p(y)) \in Y \text{ for every } x, y \in Y. \tag{1.7}$$

Clearly, Y is N-convex if and only if $N_p(Y)$ is a closed subspace of C_q. Also, Y is N-convex if and only if

$$(x^{p-1} + y^{p-1})^{1/(p-1)} \in Y \text{ for every } x, y \in Y. \tag{1.8}$$

Corollary 1.3 *A closed subspace Y of $C_p, 1 < p < \infty$, is the range of a contractive projection from C_p if and only if it is N-convex.*

Notice that Corollary 1.3 proves the equivalence of (1) and (3) in the Main Theorem.

Given a subspace Z of C_p, we define the "N-convex hull" Y of Z to be the intersection of all N-convex subspaces of C_p containing Z. Thus Y is the smallest N-convex subspace of C_p containing Z. Equivalently, Y is the smallest subspace of C_p which contains Z and is the range of a contractive projection from C_p. Y can be constructed from Z in the following way. For a subset A of C_p define $\Phi(A) = \overline{span}(A)$ and

$$\Psi(A) = \{N_q(\sum_{j=1}^{n} N_p(a_j)); n = 1, 2, \dots, a_j \in A\}.$$

Let $Z_0 = Z, W_0 = \Psi(Z_0)$ and define inductively $Z_{n+1} = \Phi(W_n)$ and $W_{n+1} = \Psi(Z_{n+1})$. Then

$$Z_0 \subseteq W_0 \subseteq Z_1 \subseteq W_1 \subseteq \dots \subseteq Z_n \subseteq W_n \subseteq \dots.$$

Thus $Y := \overline{\bigcup_{n=0}^{\infty} Z_n} = \overline{\bigcup_{n=0}^{\infty} W_n}$ is the N-convex hull of Z.

Proposition 1.4 *Let $1 < p < \infty$, and let $T : C_p \to C_p$ be an operator with $\|T\| = 1$. Then there exists a contractive projection P on C_p* with $R(P) = \ker(I - T)$ *and $PT = TP$.*

Indeed, the projection P is defined by the strong limit

$$Px = \lim_{n\to\infty} \frac{1}{n+1} \sum_{k=0}^{n} T^k x, \quad x \in C_p, \tag{1.9}$$

see [$DS1$, Chapter VIII.5]. We call this projection the *ergodic projection* associated with T and denote it by $E(T)$.

Corollary 1.5. . *The set of contractive projections in C_p for $1 < p < \infty$ is closed under infimum of an arbitrary subfamily. That is, the intersection of any family of subspaces of C_p which are ranges of contractive projections is again the range of a* *contractive projection. Moreover the infimum of two contractive projections P, Q is given by the strong limit*

$$P \wedge Q = E(PQ) = \lim_{n\to\infty} \frac{1}{n+1} \sum_{k=0}^{n} (PQ)^k \tag{1.10}$$

and is the unique contractive projection onto $R(P) \cap R(Q)$

Proof: The first statement follows from Corollary 1.3 and the fact that the intersection of a family of N-convex spaces is again N-convex. The second statement follows from Proposition 1.4 and the fact that $PQx = x$ if and only if $Px = Qx = x$. Indeed, if $PQx = x$ then we have equality in the inequalities

$$\|x\|_p = \|PQx\|_p \leq \|Qx\|_p \leq \|x\|_p.$$

Hence, $Qx = x = Px$ by Proposition 1.1. □

Remark: Let $\{P_\alpha\}_{\alpha\in A}$ be a family of contractive projections in C_p with ranges $\{X_\alpha\}_{\alpha\in A}$ and let X be the N-convex hull of $\text{span}\{X_\alpha; \alpha \in A\}$. Then X is the range of a contractive projection from C_p, which is the supremum of the family $\{P_\alpha\}_{\alpha\in A}$. With this definition of supremum and Corollary 1.5 the set of contractive projections in C_p becomes a complete lattice.

Notice that this notion of supremum is different from the ordinary one, because the closed linear span of two N-convex subspaces need not be N-convex. Here is a simple example in C_p^3, actually in l_p^3. Let

$$x = \begin{bmatrix} 1 & 0 & 0 \\ 0 & 1 & 0 \\ 0 & 0 & 0 \end{bmatrix}, \quad y = \begin{bmatrix} 0 & 0 & 0 \\ 0 & 1 & 0 \\ 0 & 0 & 1 \end{bmatrix}, \quad X = \mathbf{C}x, \ Y = \mathbf{C}y.$$

Then X, Y are N-convex as one-dimensional spaces (apply the Hahn-Banach theorem to get the contractive projections), but $X+Y=\text{span}(X \cup Y)$ is not. The reason for the last statement is that in l_p the range of a contractive projection is spanned by a family of disjointly supported elements. The space $X + Y$, considered as a subspace of l_p^3, is obviously not such a space.

Proposition 1.6. *Let P, Q be contractive projections on $C_p, 1 < p < \infty$. If $(PQ)^2$ is a projection, then PQ is also a projection. Thus $PQ = P \wedge Q$ is the contractive projection onto $R(P) \cap R(Q)$.*

Proof: Set $T = PQ$. Since T^2 is a projection, the spectrum of T is contained in $\{0, 1, -1\}$ and C_p is the direct sum $C_p = X_0 \oplus X_1 \oplus X_{-1}$ of the corresponding eigenspaces. We have to show that $X_{-1} = \{0\}$. Let $x \in X_{-1}$; then $\|x\| = \|PQx\| \leq \|Qx\| \leq \|x\|$. By Proposition 1.1, $x \in R(P) \cap R(Q)$. But then $-x = Tx = PQx = x$, so $x = 0$. □

Corollary 1.7. *Let P, Q be contractive projections on $C_p, 1 < p < \infty$. Suppose that either $PQ = QPQ$ or $PQ = PQP$. Then $PQ = QP$, and PQ is the contractive projection onto $R(P) \cap R(Q)$.*

Proof: If $PQ = QPQ$, then $(PQ)^2 = P(QPQ) = P(PQ) = PQ$. By Proposition 1.6, PQ is the contractive projection onto $R(P) \cap R(Q)$. But also $(QP)^2 = (QPQ)P = PQP$ and thus $(QP)^4 = (PQP)^2 = P(QPQ)P = P(PQ)P = PQP = (QP)^2$. By Proposition 1.6, QP is a projection. Since $\|QP\| \leq 1$, we get by Proposition 1.1 that $R(QP) = R(P) \cap R(Q)$. Thus $R(PQ) = R(QP)$, and by Proposition 1.2, $PQ = QP$.

If $PQ = PQP$, then $Q^*P^* = P^*Q^*P^*$. By duality and the first case treated above, $Q^*P^* = P^*Q^*$. Hence $PQ = (Q^*P^*)^* = (P^*Q^*)^* = QP$. □

2. JC^*-Triples and the Formulation of the Main Result

In order to formulate percisely our main result we need to introduce the C_p ideals of the classical Cartan factors. These factors are important special cases of JB^*-triples, and our results are best expressed in the JB^*-triple formalism. Therefore we shall first review some notions from the theory of JB^*-triple with the emphasis on the JC^*-triples

A JB^*-triple is a complex Banach space U, equipped with a (Jordan) triple product $U \times U \times U \to U$ denoted by $(x,y,z) \to \{x,y,z\}$, with the following properties:

(1) $\{x,y,z\}$ is bilinear and symmetric in x,z and conjugate-linear in y;

(2) $\|\{x,y,z\}\| \leq \|x\|\|y\|\|z\|$, for any $x,y,z \in U$;

(3) for every $a \in U$, the operator $D(a)$ defined by $D(a)x = \{a,a,x\}$ is Hermitian (i.e. $\|e^{itD(a)}\| = 1, \quad t \in \mathbf{R}$), its spectrum is non-negative and $\|D(a)\| = \|a\|^2$;

(4) for every $a \in U$, the operator $\delta(a) := iD(a)$ is a *derivation* of the triple product:

$$\delta(a)\{x,y,z\} = \{\delta(a)x,y,z\} + \{x,\delta(a)y,z\} + \{x,y,\delta(a)z\}.$$

By polarizing the last identity, one obtains the so called *Main Identity of the triple product*

$$\{a,b,\{x,y,z\}\} = \{\{a,b,x\},y,z\} - \{x,\{b,a,y\},z\} + \{x,y,\{a,b,z\}\}. \tag{2.1}$$

The main subclass of JB^*-triples are the JC^*-triples, defined as follows: A *JC^*-triple* is a norm closed subspace U of $B(H)$ such that $xx^*x \in U$ whenever $x \in U$. This notion was introduced by L. Harris $[H]$, where the terminology "J^*-algebras" was used. By polarization, we see that a closed subspace U of $B(H)$ is a JC^*-triple if and only if it is closed under the *triple product*

$$\{x,y,z\} := (xy^*z + zy^*x)/2. \tag{2.2}$$

The JC^*-triples are precisely the JB^*-subtriples of $B(H)$ for some Hilbert space H, i.e., properties (1) - (4) of the above definition hold for the triple product (2.2). A *JCW^*-triple* is a JC^*-triple which is closed in the w^*-topology of $B(H)$. The JCW^*-triples are precisely the JC^*-triples which are dual Banach spaces. Similarly, the JBW^*-triples are the JB^*-triples which are dual spaces. The general references for JB^*-triples are $[U1], [U2], [K1]$ and $[Lo]$.

We use below the triple product formalism in the JC^*-triples (rather than the binary product) not just for convention and simplification, but mainly because it reflects most clearly the underlying geometry.

A linear mapping T from a JB^*-triple U into a JB^*-triple V is a *triple homomorphism* if for all $x, y, z \in U$

$$T\{x, y, z\} = \{Tx, Ty, Tz\}. \tag{2.3}$$

A triple homomorphism T is a *triple monomorphism* if it is one-to-one, and a *triple isomorphism* if it is one-to-one and onto. Using the "C^* *condition*" in JB^*-triples

$$\|\{x, x, x\}\| = \|x\|^3, \qquad x \in U \tag{2.4}$$

it is not hard to see that a triple monomorphism is in fact an isometry. Conversely, it is known that an isometry form one JB^*-triple U onto another is necessarily a triple-isomorphism.

The building blocks of the algebraic structure of JB^*-triples are the *tripotents* ("triple idempotents"), i.e. elements $u \in U$ satisfying $\{u, u, u\} = u$. Clearly, in a JC^*-triple the tripotents are precisely the partial isometries . Two tripotents u, v are *orthogonal* if $u + v$ and $u - v$ are tripotents. The notation for orthogonality is $u \perp v$. Clearly, if u and v are tripotents in a JC^*-triple, then $u \perp v$ if and only if their initial and final spaces are orthogonal, i.e.

$$uv^* = 0 = u^*v. \tag{2.5}$$

In general $u \perp v$ is equivalent to either $\{u, u, v\} = 0$ or to $\{v, v, u\} = 0$. Thus a triple-isomorphism maps orthogonal tripotents into orthogonal tripotents.

On the set of tripotents in the JB^*-triple U we define a *parial ordering* as follows: $u \leq v$ if and only if $v = u + w$ where w is a tripotent in U and $w \perp u$. A tripotent u is *minimal* in U if for any tripotent $v \in U$, $v \leq u$ implies either $v = 0$ or $v = u$.

To each element x in the JB^*-triple U we associate an operator $Q(x) : U \to U$, via

$$Q(x)y = \{x, y, x\}; \quad y \in U. \tag{2.6}$$

$Q(x)$ is conjugate-linear, and

$$\|Q(x)\| = \|x\|^2. \tag{2.7}$$

If $v \in U$ is a tripotent, then the operator $D(v) = \{v, v, \cdot\}$ satisfies:

$$D(v)(D(v) - I)(2D(v) - I) = 0. \tag{2.8}$$

Thus the spectrum of $D(v)$ consists of eigenvalues only and it is contained in $\{0, 1/2, 1\}$. Let $U_k(v)$ be the eigenspace of $D(v)$ corresponding to the eigenvalue $k/2$, and let $P_k(v)$ be the corresponding spectral projection onto $U_k(v)$, $k = 0, 1, 2$. $U_k(v)$ are the *Peirce subspaces* of U associated with v, and $P_k(v)$ are the corresponding *Peirce projections* of v. Clearly, the $P_k(v)$ can be expressed as polynomials in $D(v)$, and $P_2(v) = Q(v)^2$.

Since for every $x \in U$, $\delta(x) := iD(x)$ is a triple derivation, it follows that for any tripotent $v \in U$,

$$\{U_j(v), U_k(v), U_l(v)\} \subseteq U_{j-k+l}(v) \tag{2.9}$$

where $U_m(v) = \{0\}$ if $m \notin \{0, 1, 2\}$. The multiplicatoin rules (2.9) are called the *Peirce calculus.*

Moreover,

$$\{U_0(v), U_2(v), U\} = \{U_2(v), U_0(v), U\} = \{0\} \tag{2.10}$$

and each of the spaces $U_k(v)$ is a JB^*-subtriple, $k = 0, 1, 2$.

The *Peirce reflection* associated with a tripotent $v \in U$ is the operator $S(v) = P_2(v) - P_1(v) + P_0(v) = \exp(2\pi i D(v))$. It is the symmetry ($S^2(v) = I$) fixing elementwise $U_2(v) \oplus U_0(v)$ and satisfying $S(v)|_{U_1(v)} = -I|_{U_1(v)}$. The operator $Q(v)$ is a conjugate-linear triple automorphism of $U_2(v)$ of period 2, and is used to define there an *involution* via

$$x^\# := Q(v)x = \{v, x, v\}, \quad x \in U_2(v). \tag{2.11}$$

In case of a JC^*-triple we get $x^\# = vx^*v$. This involution gives rise to relative (i.e. with respect to v) selfadjoint elements of $U_2(v)$. An operator $T : U \to U$ is *locally* *self-adjoint* if $TQ(v) = Q(v)T$ for every tripotent $v \in U$. Since $Q(v)^2 = P_2(v)$, this implies that $T(U_2(v)) \subseteq U_2(v)$, and that $T|_{U_2(v)}$ is selfadjoint with respect to the involution (2.11).

Two tripotents $v, w \in U$ are *compatible* if

$$P_j(v)P_k(w) = P_k(w)P_j(v) \quad \text{for all } j,k \in \{0,1,2\}. \tag{2.12}$$

By $[Mc], v$ and w are compatible if and only if

$$P_j(v)w \in U_2(w) \quad \text{for all } j = 0,1,2. \tag{2.13}$$

In particular, if $w \in U_k(v)$ for some $k \in \{0,1,2\}$ then v and w are compatible.

Given a sequence $\{v_j\}_{j=1}^n$ $(n \leq \infty)$ of orthogonal tripotents in U we define the associated *joint Peirce projections* $\{P_{i,j}\}_{0 \leq i \leq j \leq n}$ by

$$\begin{aligned} P_{i,i} &:= P_2(v_i), \quad 1 \leq i \leq n; \\ P_{i,j} &:= P_1(v_i)P_1(v_j), \quad 1 \leq i < j \leq n; \\ P_{0,j} &:= P_1(v_j) \prod_{1 \leq i \leq n, i \neq j} P_0(v_i), \quad 1 \leq j \leq n; \\ P_{0,0} &:= \prod_{1 \leq i \leq n} P_0(v_i). \end{aligned} \tag{2.14}$$

The *joint Peirce decomposition* of U is

$$U = \sum_{0 \leq i \leq j \leq n} \oplus U_{i,j}, \quad \text{where } U_{i,j} = P_{i,j}U. \tag{2.15}$$

We will need some results on the triple functional calculus. Since the general JB^*-triple case involves some complications, we assume from this point on that U is a JC^*-subtriple of $B(H)$. This suffices for our purposes.

Given $x \in U$ with a polar decomposition $x = v|x|$ and a continuous function f on $[0,\infty)$ we define

$$f(x) = vf(|x|), \tag{2.16}$$

where $f(|x|)$ is defined by the ordinary functional calculus. If $f(0) = 0$ and if f is extended to an anti-symmetric function on $\mathbf{R}$ then for every self adjoint element

x, $f(x)$ defined by (2.16) coincides with $f(x)$ defined via ordinary functional calculus, see [ABF]. We call the map $f \mapsto f(x)$ the *triple functional calculus* associated with x. See section 3 for more details. It is known that $f(x) \in U$, provided $f(0) = 0$ in case $0 \in \sigma(|x|)$.

The *singular numbers* $\{s_n\}$ of a compact operator x on H are the eigenvalues of $|x|$, arranged in a non-increasing ordering, counting multiplcity. By the spectral theorem and polar decomposition, x admits a series expansion of the form $x = \sum_n s_n(\cdot, e_n)f_n$ where $\{e_n\}$ and $\{f_n\}$ are orthonormal sequences, consisting of eigenvectors of $|x|$ and $|x^*|$ respectively. This expansion is called in [GK1] the *Schmidt series* of x. Let $\{\alpha_j\}$ be the distinct singular numbers of x arranged in a strictly decreasing ordering. Then the Schmidt series of x takes the form

$$x = \sum_j \alpha_j v_j, \tag{2.17}$$

where $\{v_j\}$ are orthogonal finite-rank tripotents (i.e., partial isometries). In what follows we call (2.17) the Schmidt series of x. Notice that (2.17) has a meaning in a general JB^*-triple. If f is a continuous function on $[0, \|x\|]$, then

$$f(x) = \sum_j f(\alpha_j)v_j. \tag{2.18}$$

For any JC^*-triple U and $1 < p < \infty$ we define

$$U_p = U(C_p) = U \cap C_p. \tag{2.19}$$

If $x \in U_p$ is given by (2.17) then

$$N_p(x) = x^{p-1}/\|x\|_p^{p-2} = \sum_j \alpha_j^{p-1} v_j/\|x\|_p^{p-2} \in U_q, \tag{2.20}$$

where $q^{-1} + p^{-1} = 1$. Thus

$$N_p(U_p) = U_q, \quad N_q(U_q) = U_p \tag{2.21}$$

Given $x \in U$ with a polar decomposition $x = v|x|$, we define the *Peirce projections of* x by $P_k(x) = P_k(v), k = 0, 1, 2$. Notice that if $r(x) := v^*v$ and $\ell(x) := vv^*$

are the *right and left support projections of* x, then

$$P_2(x)y = \ell(x)yr(x), \ P_1(x)y = \ell(x)y(I - r(x)) + (I - \ell(x))yr(x),$$

and

$$P_0(x)y = (I - \ell(x))y(I - r(x)).$$

If $y \in U$ has a polar decomposition $y = w|y|$, we say that x and y are *orthogonal* (and write $x \perp y$) if $v \perp w$. Clearly, $x \perp y$ if and only if $\ell(x)\ell(y) = 0 = r(x)r(y)$. Also, $x \perp y$ if and only if $x^*y = 0 = xy^*$. Two subsets X, Y of U are *orthogonal*, notation $X \perp Y$, if $x \perp y$ for every $x \in X$, $y \in Y$. Orthogonal direct sum in C_p is the same as the direct sum in the sense of ℓ_p, $1 \leq p < \infty, p \neq 2$. Namely, subsets X, Y are orthogonal if and only if $\|x + y\| = (\|x\|^p + \|y\|^p)^{1/p}$ for every $x \in X$ and $y \in Y$. This follows from the equality case in Clarkson inequality (See $[M], [AF1], [Y]$).

If $x \perp y$ and f is a continuous function on $[0, \infty)$ then

$$f(x) \perp f(y) \text{ and } f(x + y) = f(x) + f(y). \tag{2.22}$$

From this the following proposition follows easily.

Proposition 2.1 *If* X, $Y \subset C_p$ $(1 < p < \infty)$ *are orthogonal subspaces which are* N*-convex, then* $N_p(X), N_p(Y)$ *are orthogonal subspaces of* C_q $(q^{-1} + p^{-1} = 1)$ *and* $N_p(X + Y) = N_p(X) + N_p(Y)$. *Thus* $X + Y$ *is also* N*-convex. Conversely, if* Z *is an* N*-convex subspace of* C_p *and* $Z = X + Y$ *where* X, Y *are orthogonal subspaces, then each of* X, Y *is* N*-convex.*

A subspace Z of C_p or $B(H)$ is *decomposable* if it is the sum of two non-trivial orthogonal subspaces. Otherwise, Z is *indecomposable*. By Zorn's lemma, one gets

Proposition 2.2 *Every subspace* Z *of* C_p, $1 < p < \infty$, *is the orthogonal direct sum* $Z = \sum_k \oplus Z_k$ *of a family of indecomposable subspaces* Z_k. *Moreover,* Z *is* N*-convex if and only if every* Z_k *is* N*-convex.*

Given Hilbert spaces H, K we form the Hilbert space tensor product $H \otimes K$. It is known that $B(H \otimes K)$ is the C^*-tensor product $B(H) \otimes B(K)$. Also $C_p(H \otimes K) =$

$C_p(H) \otimes C_p(K)$, namely the linear span of $x \otimes y, x \in C_p(H), y \in C_p(K)$, is dense in $C_p(H \otimes K)$. Clearly, $\|x \otimes y\|_p = \|x\|_p \|y\|_p$ and $N_p(x \otimes y) = N_p(x) \otimes N_p(y)$. Also, if $x_1 \perp x_2$ then $(x_1 \otimes y) \perp (x_2 \otimes y)$ for all y.

We use tensor products mainly for notational convenience. For instance, two elements x and y of $B(H)$ are orthogonal if and only if there is a realization of H as $H \otimes H$, and corresponding realizations of $B(H)$ as $B(H) \otimes B(H)$ so that $x = e_{1,1} \otimes a$ and $y = e_{2,2} \otimes b$ for some $a, b \in B(H)$. Every particular realization of $B(H)$ as $B(H) \otimes B(H)$ (and of C_p as $C_p \otimes C_p$) is called a *tensor product representation.* See [AF2] for more details.

A JC^*-triple (resp. JCW^*-triple) U is *atomic* if it is the norm (resp. w^*-) closed linear span of its minimal tripotents. A tripotent $v \in U$ is minimal if and only if $\dim U_2(v) = 1$ i.e. v is an "atom" of U. This follows from the fact that $U_2(v)$ is a JB^*-algebra with a unit v, and thus

$$U_2(v) = \overline{\text{span}}\{w;\ w \in U,\ w \text{ is a tripotent, and } w \leq v\}.$$

If U is indecomposable, then U is atomic if and only if it contains a minimal tripotent.

The indecomposable JC^*-triples are called *factors.* Atomic JCW^*-factors are known as *Cartan factors* of types I-IV, or the "classical" (or "spatial") Cartan factors. (The exceptional Cartan factors of types V and VI have dimensions 16 and 27 respectively and are not subtripes of $B(H)$).

The classification of classical Cartan factors up to a triple-isomorphism is the following (n, m are arbitrary cardinal numbers).

Type $I_{n,m}$: $U = B(H, K)$, where H and K are Hilbert spaces of dimensions n and m respectively;

Type II_n : $U = \{x \in B(H); x^T = -x\}$, where x^T is the transpose of x relative to a fixed orthonormal basis and $\dim(H) = n$;

Type $III_n : U = \{x \in B(H); x^T = x\}$, notation as for type II_n;

Type $IV_n : U$ is any subspace of $B(H)$ of dimension n, so that for any $x \in U, x^* \in U$ and $x^2 \in \mathbf{C}I$.

We denote these Cartan factors by $U(I_{n,m})$, $U(II_n)$, $U(III_n)$ and $U(IV_n)$ respectively. There are some coincidences of the classical Cartan factors in low dimensions, see [Lo].

The classical Cartan factors are spanned by canonical systems of tripotents, called "*colinear-ortho-governing grids*" (or, *cog*-grids, for short) in [MM], [N] and [DF] (The relations of "colinearity" and "governing" will be defined in Section 6 below). Let n, m be cardinal numbers and realize $U(I_{n,m})$ as $B(H, K)$ where H, K are Hilbert spaces of dimensions n and m respectively. Choose orthonormal bases $\{f_j\}_{j\in J}$ for H and $\{e_i\}_{i\in I}$ for K, and let $e_{i,j} = (\cdot, f_j)e_j$ $(i \in I, j \in J)$ be the corresponding *matrix units*. Then $e_{i,j}$ are minimal tripotents in $B(H, K)$ and $B(H, K) = \overline{span}^{w^*}\{e_{i,j}\}_{j\in J, i\in I}$, where the closure is taken in the w^*-topology. $\{e_{i,j}\}_{i\in I, j\in J}$ is a *cog*-grid for $B(H, K)$ and every other *cog*-grid has the same form with respect to a pair of orthonormal bases. If $H = K$ and the index set $I = J$ is well-ordered by $\leq$, we define $a_{i,j} = e_{i,j} - e_{j,i}, s_{i,j} = e_{i,j} + e_{j,i}$; $i < j$. Then the canonical *cog*-grids in $U(II_n)$ and $U(III_n)$ are $\{a_{i,j}; i, j \in I, i < j\}$ and $\{s_{i,j}; i, j \in I, i < j\} \cup \{e_{i,i}; i \in I\}$ respectively. Again, the *cog*-grids are unique up to the choice of the orthonormal bases $\{e_i\}_{i\in I}, \{f_i\}_{i\in I}$.

The Cartan factor $U(IV_n)$ is called the *spin factor*. Its canonical cog-grid is called a *spin grid* and is defined as follows (see [*DF* : *p*.313]). If n is an odd number, then U is the span of a family of tripotents $\{u_j, \tilde{u}_j\}_{j=1}^{(n-1)/2} \cup \{u_0\}$ such that for any $1 \leq i, j \leq (n-1)/2$, $i \neq j$ there is a triple monomorphism of $U(I_{2,2})$ into $U(IV_n)$ mapping

$$e_{1,1} \to u_i,\ e_{1,2} \to u_j,\ e_{2,2} \to \tilde{u}_i,\ e_{2,1} \to -\tilde{u}_j, \tag{2.23}$$

and for any $1 \leq j \leq (n-1)/2$ there is a triple monomorphism of $U(III_2)$ into $U(IV_n)$ mapping

$$e_{1,1} \to u_j,\ e_{1,2} + e_{2,1} \to u_0,\ e_{2,2} \to -\tilde{u}_j. \tag{2.24}$$

For all other cardinal numbers n, $U(IV_n)$ is the norm closed span of a family $\{u_j, \tilde{u}_j\}_{j\in J}$ satisfying (2.23) with $n = |J| + |J|$.

Let U be a Cartan subfactor of $B(H)$ and let $\{u_j\}_{j\in J}$ be a cog-grid in U. Then $Ex = \sum_{j\in J}\langle x, u_j\rangle u_j/\|u_j\|_2^2$ is the orthonormal projection from C_2 onto U_2. It is a fundamental fact that E extends to a contractive projection from $B(H)$ onto U. Since E is self-adjoint with respect to the C_2-inner product, it follows that it is a contractive projection from C_1 onto U_1. By interpolation (see [A3]) or direct computations as in [AF2], it follows that E is a contractive projection from C_p onto U_p. The canonical contractive projections arising in this way are called *conditional expectations*. They wil be generalized in Theorem 8.17 and Remark 8.18. If U is of one of the types I-III, then the conditional expectation E has a very explicit form. If $U = U(I_{n,m})$, then $Ex = pxq$, where p and q are projections on H whose ranges have dimensions n and m respectively. If $U = U(II_n)$ (or $U = U(III_n)$), then $Ex = p[(x - x^T)/2]p$ (respectively, $Ex = p[(x + x^T)/2]p$), where p is as before.

We will need some more detailed information on the structure of atomic JCW^*-triples. By [Ho], [DF] every atomic JBW^*-triple is an orthogonal direct sum of a family of indecomposable ideals, and every indecomposable atomic JBW^*-triple is triple isomorphic to a Cartan factor.

A triple-homomorphism T of a JC^*-triple U into $B(H)$ will be called a *triple-representation* (or, *representation*, for short). T is *faithful* if it is one-to-one (i.e., monomorphism); in this case it is necessarily an isometry. Assume now that U is a classical Cartan factor and that $T : U \to B(H)$ is a faithful representation. T is *reducible* if there exist faithful representations $T_1, T_2 : U \to B(H)$ whose ranges are orthogonal, so that $T = T_1 + T_2$. Otherwise T is *irreducible*. Two faithful representations $T_j : U \to B(H)$ $(j = 1, 2)$ are *equivalent* if there exist partial isometries $v, w \in B(H)$ so that $T_2(x) = uT_1(x)w$ for all $x \in U$. Otherwise T_1, T_2 are *inequivalent.*

By Zorn's lemma, it is easy to see that every faithful representation $T : U \to B(H)$ can be written in a tensor product notation as

$$Tx = \sum_{j \in J} v_j \otimes T_j(x), \quad x \in U,$$

where $\{T_j\}_{j \in J}$ are mutually inequivalent irreducible faithful representations and $\{v_j\}_{j \in J}$ are pairwise orthogonal tripotents.

The classification of the irreducible faithful representations of classical Cartan factors is obtained in [AF2]. Up to equivalence, the factors $U(II_n), U(III_n)$ and $U(IV_n)$ for n odd admit only one irreducible faithful representation (the identity map). They correspond in [AF2] to A^n_∞, SY^n_∞ and DAH^n_∞ respectively. The factors $U(I_{n,m})$ for $\min\{n, m\} \geq 2$ and $U(IV_n)$ for n even admit two inequivalent faithful representations: the identity map and the transposition. These factors correspond in $[AF2]$ to $C^{n,m}_\infty$ and AH^n_∞ respectively. The Hilbert spaces $U(I_{1,n})$ (corresponding to the spaces $H^n_\infty(\{y_j\}^n_{j=1})$ in [AF2]) admit precisely $n+1$ inequivalent, irreducible faithful representations. These representations are described in full detail in [AF2] and are fairly complicated. Their exact structure will be needed only in Section 7.

Let U be a Cartan factor, let u be a minimal tripotent in U and let $T : U \to B(H)$ be a triple monomorphism. Then Tu is a tripotent of rank r, and

$$\|Tx\|_p = (r)^{1/p} \|x\|_p; \quad x \in U_p.$$

We denote $\alpha(T) = r$.

From these considerations the following is easily obtained.

Proposition 2.3 *Let U be a classical Cartan factor, and let H, K be Hilbert spaces. Let $\{a_j\}_{j \in J}$ be an orthogonal family in $C_p(K)$, $1 < p < \infty$, and let T_j: $U \to B(H), j \in J$, be faithful representations. Assume that*

$$\sum_{j \in J} \|a_j\|_p^p \alpha(T_j)^p = 1. \tag{2.25}$$

Let

$$Ty = \sum_{j \in J} a_j \otimes T_j(y), \qquad y \in U_p. \tag{2.26}$$

Then T is an isometry from U_p into $C_p(K \otimes H) \equiv C_p(K) \otimes C_p(H)$, and $X = T(U_p)$ is an N-convex subspace of $C_p(K \otimes H)$. Moreover X is indecomposable.

Operators $T : U_p \to C_p \equiv C_p(K \otimes H)$ of the form (2.26) will be called *triple-isometries.* We discuss their properties in greater detail and generality in section 8 below. We would like to mention only one fact concering the triple isometries: if the rank of U is ≥ 2, then every isometry of U_p into C_p, $1 \leq p < \infty$, $p \neq 2$, is in fact a triple isometry, and therefore there is a contractive projecion from C_p onto its range. Proposition 2.3 can be restated now as follows: The range of a triple isometry $T : U_p \to C_p$ is the range of a contractive projection from C_p.

The most difficult part of the Main Theorem is the converse of Proposition 2.3, namely, the implication (1) $\Rightarrow$ (2). This is formulated in the following theorem.

Theorem 2.4 *Let $1 < p < \infty, p \neq 2$, and let X be an indecomposable subspace of C_p which is the range of a contractive projection from C_p. Then there exists a classical Cartan factor U, an orthogonal family $\{a_j\}_{j \in J}$ in C_p and a family of faithful triple-representations T_j: $U \to B(H), j \in J$, so that*

$$\sum_{j \in J} \|a_j\|_p^p \alpha(T_j)^p = 1,$$

and

$$X = \left\{ \sum_{j \in J} a_j \otimes T_j(y); y \in U_p \right\}, \tag{2.27}$$

where $U_p = U(C_p)$ is the C_p ideal of the Cartan factor U. Thus the operator T defined by (2.26) is a triple-isometry of U_p onto X.

Combining Propositions 2.2, 2.3 and Theorem 2.4 we get the following:

Corollary 2.5 *Let $1 < p < \infty, p \neq 2$, and let X be a subspace of $C_p = C_p(H)$. Then the following are equivalent:*

(a) *X is the range of a contractive projection from C_p;*

(b) *X is the ℓ_p-sum of subspaces of C_p, each of which is triple isometric to the C_p ideal of a classical Cartan factor.*

Notice that Corollary 2.5 is the equivalence (1) $\Leftrightarrow$ (2) in the Main Theorem.

3. Differentiation formulas and Schur multipliers

Let J be a finite or infinite interval in $\mathbf{R}$. Let $C^1(J)$ be the space of continuously differentiable functions with bounded derivatives on J. For $0 < \epsilon < 1$ let $C^{1+\epsilon}(J)$ denote the space of all functions f in $C^1(J)$, for which $f' \in Lip_\epsilon(J)$, with the semi norm $\|f\|_{C^{1+\epsilon}(J)} = \|f'\|_{Lip_\epsilon(J)}$. The *divided differences* of $f \in C^1(J)$ are denoted by

$$f^{[1]}(s,t) = \begin{cases} \dfrac{f(s)-f(t)}{s-t}; & s \neq t \\ f'(t); & s = t \end{cases} \tag{3.1}$$

Note that $f^{[1]}$ is bounded and continuous on $J \times J$.

Let $B(H)_{s.a.}$ denote the space of self-adjoint operators in $B(H)$. Similarly, $(C_p)_{s.a} = C_p \cap B(H)_{s.a}$. Let $a \in B(H)_{s.a.}$ with spectrum $\sigma(a) \subset J$ and spectral decomposition $a = \int_{\mathbf{R}} \lambda de_\lambda$, where the spectral measure $\{e_\lambda\}_{\lambda \in \mathbf{R}}$ is constant outside J. For every continuous function f on J define $f(a) = \int_{\mathbf{R}} f(\lambda) de_\lambda$. The *functional calculus* $f \mapsto f(a)$ is well behaved. We are interested here in the study of the smoothness properties of the map $a \mapsto f(a)$. The following Theorem summarizes several results from [BSIII].

Theorem 3.1. *Let $0 < \epsilon < 1$ and let $f \in C^{1+\epsilon}(J)$. Let $a, b \in B(H)_{s.a}$, assume that $\sigma(a) \subset J$ and let $a = \int_{\mathbf{R}} \lambda de_\lambda$ be the spectral decomposition of a.*

(i) *The integral*

$$F(a;b) = \int_{\mathbf{R}^2} f^{[1]}(\lambda_0, \lambda_1) de_{\lambda_0} b \, de_{\lambda_1} \tag{3.2}$$

converges in the operator norm, and

$$\|F(a;b)\| \leq C(J,\epsilon)\|f\|_{C^{1+\epsilon}(J)}\|b\|. \tag{3.3}$$

(ii) *The function $t \to f(a+tb)$ is continuously differentiable in the operator norm in a neighborhood of $t_0 = 0$, and*

$$\frac{d}{dt}f(a+tb)_{|_{t=0}} = F(a;b). \tag{3.4}$$

(iii) *If $b \in (C_p)_{s.a.}, 1 \leq p \leq \infty$, then the integral (3.2) converges in C_p,*

$$\|F(A;b)\|_p \leq C(J,\epsilon)\|f\|_{C^{1+\epsilon}(J)}\|b\|_p, \tag{3.5}$$

and the differentiation in (3.4) is in the norm of C_p.

Parts (i) and (ii) of Theorem 3.1 are proved in [DK] under stronger assumptions on f. The theory of the *Stieltjes double operator integrals* (3.2) is developed in [BSI] and [BSII]. In [P] the condition $f \in C^{1+\epsilon}(J)$ is replaced by a weaker one (the membership of f in an appropriate Besov space), which is shown to be almost optimal. The papers [F1], [F2], [F3], [A2] and [ABF] contain more results concerning the differentiability of the map $a \mapsto f(a)$. A function $f \in C^1(J)$ is said to be *operator differentiable* if parts (i), (ii) of Theorem 3.1 hold for every $a, b \in B(H)_{s.a.}$ with $\sigma(a) \subset J$.

Motivated by the *Schur product* of matrices (sometimes called also Hadamard product)

$$(a_{i,j}) \circ (b_{i,j}) = (a_{i,j}b_{i,j}), \tag{3.6}$$

we shall call the integral operators of the form

$$\Phi(x) = \int\int_{\mathbf{R}^2} \varphi(s,t)de_s\, x\, df_t \tag{3.7}$$

Schur multipliers. Here $\{e_s\}_{s\in\mathbf{R}}$ and $\{f_t\}_{t\in\mathbf{R}}$ are spectral measures and $\varphi(s,t)$ a measurable function. See [P] for the characterization of those functions φ for which

Φ is a bounded Schur multiplier. Thus the operator $b \to F(a;b)$ is the Schur multiplier corresponding to $\varphi = f^{[1]}$.

If $a \in B(A)_{s.a.}$ has a discrete spectral decomposition $a = \sum_j \lambda_i e_j$ with $\{\lambda_j\}$ distinct eigenvalues corresponding to the spectral projections $\{e_j\}$, then using the fact that $f^{[1]}(s,t) = f^{[1]}(t,s)$, (3.2) becomes

$$F(a;b) = \sum_{i,j} f^{[1]}(\lambda_i,\lambda_j) e_i b e_j = \sum_{0 \le i \le j} f^{[1]}(\lambda_i,\lambda_j) P_{i,j} b, \tag{3.8}$$

where $\{P_{i,j}\}$ are the joint Peirce projections associated with $\{e_j\}$, see (2.14).

We will need in Sections 4 and 5 some information on the smoothness of the support functional map $N_p : C_p \to C_q$ $(1 < p < \infty, \frac{1}{p} + \frac{1}{q} = 1)$. The natural framework for this study is the *triple functional calculus* defined by (2.16) and (2.18).

Let U be a JC^*-subtriple of C_∞ and fix $a \in U$ with a Schmidt series $a = \sum_j \alpha_j v_j$, where $\alpha_1 > \alpha_2 > \cdots > \alpha_j \cdots > 0$ and $\{v_j\}$ are orthogonal, finite-rank tripotents from U. Let $v = \sum_j v_j$ and let $Q(v)x = \{vxv\} = vx^*v$. Then $Q(v)^2 = P_2(v)$ and $Q_2(v)|_{U_2(v)}$ is an involution. Define *symmetrization* and *anti-symmetrization* operators S_v and A_v respectively by

$$S_v = (I + Q(v))/2, \; A_v = (I - Q(v))/2. \tag{3.9}$$

Then $S_v + A_v = I, S_v - A_v = Q(v)$ and the operators $S_v|_{U_2(v)}, A_v|_{U_2(v)}$ are projections. Since $Q(v)$ is conjugate-linear, S_v and A_v are only real-linear and

$$S_v(ix) = iA_v(x), \quad A_v(ix) = iS_v(x). \tag{3.10}$$

Let $\{P_{i,j}\}_{0 \le i \le j}$ be the *joint Peirce projections* associated with $\{v_j\}$ via (2.14). An operator M on U is a *(triple) Schur multiplier* (relative to $\{P_{i,j}\}$) if $M = \sum_{0 \le i \le j} m_{i,j} P_{i,j}$ for an appropriate "triangle" $\{m_{i,j}\}_{0 \le i \le j}$ of scalars. Given an odd function $f \in C^1(\mathbf{R})$, we define

$$m_f^{\pm}(s,t) = \frac{f(s) \pm f(t)}{s \pm t}; \quad s,t \in \mathbf{R}, \tag{3.11}$$

where $m_f^-(t,t) = f'(t)$ and $m_f^{\pm}(0,0) = f'(0)$. We associate with f and the element $a \in U$ two Schur multipliers

$$M_{f,a}^{\pm} = \sum_{0 \le i \le j} m_f^{\pm}(\alpha_i, \alpha_j) P_{i,j}. \tag{3.12}$$

The following result is a consequence of Theorem 3.1 and [ABF,Th.3.3].

Theorem 3.2. *Let $0 < \epsilon < 1$ and let f be an odd function in $C^{1+\epsilon}(\mathbf{R})$. Let $a = \sum_j \alpha_j v_j \in C_\infty$ and $v = \sum_j v_j$ be as above.*

(i) *For every $b \in C_\infty$,*

$$\frac{d}{dt} f(a+tb)_{|t=0} = (M_{f,a}^- S_v + M_{f,a}^+ A_v) b. \tag{3.14}$$

Here the differentiation and the convergence of the series defining $M_{f,a}^- S_v b$ and $M_{f,a}^+ A_v b$ are in the operator norm, and $f'(a) := M_{f,a}^- S_v + M_{f,a}^+ A_v$ is a bounded, real-linear operator on C_∞.

(ii) *If $b \in C_p, 1 \le p < \infty$, then the differentiation and the convergence of the series are in the norm of C_p, and $f'(a)$ is a bounded, real-linear operator on C_p.*

(iii) *If U is a subtriple of C_∞ and $a \in U$, then $f'(a)$ maps U into itself and $U_p = U \cap C_p$ into itself.*

The odd functions $f \in C^1(\mathbf{R})$ for which part (i) of Theorem 3.2 holds for every $a \in C_\infty$, may be called C_∞-(triple) differentiable functions. In [ABF] we study U-(triple) differentiable functions for a general JC^*-triple, and show that they coincide with the odd operator-differentiable functions on $\mathbf{R}$.

We need to strengthen the results concerning the boundness of the Schur multipliers in (3.8) and (3.14). To illustrate one of the main ideas, consider first ordinary multipliers. If $a \in B(H)$ and $b \in C_p$ then $ab \in C_p$ and $\|ab\|_p \le \|a\|_\infty \|b\|_p$. But if $a \in C_r, 0 < r < \infty$, then ab belongs to the smaller ideal C_s $(\frac{1}{s} = \frac{1}{r} + \frac{1}{p})$ and

$\|ab\|_s \leq \|a\|_r \|a\|_p$. We will show below that if $a \in C_r$ with $r < \infty$, then the operators (3.8) and (3.14) corresponding to the function $f(\lambda) = sgn(\lambda)|\lambda|^{1+\tau}$ $(0 < \tau)$ map C_{p_1} into C_{p_2}, where $\frac{1}{p_2} = \frac{1}{p_1} + \frac{\tau}{r}$.

Next, we need some results from the theory of *integration with respect to a spectral family of projections*, taken from [Do, Chapter 5] and [BG]. A *spectral family* of projections on a Banach space X is a one-parameter family of projections $\mathcal{E} = \{E_\lambda\}_{\lambda \in \mathbf{R}}$ so that

(i) $E_{\lambda_1} E_{\lambda_2} = E_{\min\{\lambda_1, \lambda_2\}}$;

(ii) $\|\mathcal{E}\| := \sup_{\lambda \in \mathbf{R}} \|E_\lambda\| < \infty$;

(iii) The one-sided limits $\lim_{\lambda \to \lambda_0 \pm 0} E_\lambda x$ exist in the norm for every $x \in X$, and $\lim_{\lambda \to \lambda_0 + 0} E_\lambda x = E_{\lambda_0} x$;

(iv) $\lim_{\lambda \to \infty} E_\lambda x = x$ and $\lim_{x \to -\infty} E_\lambda x = 0$.

The family $\{E_\lambda\}_{\lambda \in \mathbf{R}}$ is *supported in* $[\alpha, \beta]$ if E_λ is constant on $(-\infty, \alpha)$ and on (β, ∞). The major difference between a spectral family and a spectral measure is that if $\{E_\lambda\}_{\lambda \in \mathbf{R}}$ is a spectral family and $\{\lambda_k\}_{k \in \mathbf{Z}}$ is a monotone two-sided sequence with $\lim_{k \to \infty} \lambda_k = \infty$ and $\lim_{k \to -\infty} \lambda_k = -\infty$, then for $x \in X$ the convergence of the series $x = \sum_{k=-\infty}^{\infty} (E_{\lambda_k} - E_{\lambda_{k-1}})x$ need not be unconditional. Nevertheless, the theory of integration with respect to a spectral family of projections gives quite satisfactory results.

We denote by $BV[\alpha, \beta]$ the space of all right continuous functions of bounded variation on thc compact interval $[\alpha, \beta]$, with the norm

$$\|f\|_{BV[\alpha,\beta]} = \mathrm{Var}(f; [\alpha, \beta]) + |f(\beta)|. \tag{3.15}$$

Let $\mathcal{E} = \{E_\lambda\}_{\lambda \in \mathbf{R}}$ be a spectral family of projections on the Banach space X and let $f \in BV[\alpha, \beta]$. For every $x \in X$ and a partition $\alpha = \lambda_0 < \lambda_1 < \cdots < \lambda_n = \beta$ consider the sum $\sum_{j=1}^{n} f(\lambda_j)(E_{\lambda_j} - E_{\lambda_{j-1}})x + f(\alpha) E_\alpha x$. The set of partitions is ordered by refinement. The net of the above sums converges in norm to an elemen of X denoted by $\int_{\mathbf{R}} f(\lambda)dE_\lambda x$. The operator $\int_{\mathbf{R}} f(\lambda)dE_\lambda$ on X constructed in this way is bounded and

$$\|\int_{\mathbf{R}} f(\lambda)dE_\lambda\| \leq \|f\|_{BV[\alpha,\beta]}\|\mathcal{E}\|. \tag{3.16}$$

The operators $T \in B(X)$ which can be written in the form $T = \int_{\mathbf{R}} \lambda dE_\lambda$ for some spectral family of projections $\{E_\lambda\}_{\lambda\in\mathbf{R}}$ on X which is supported in some interval $[\alpha, \beta]$ are called *well bounded operators of type (B)*. The spectral family $\{E_\lambda\}_{\lambda\in\mathbf{R}}$ is then uniquely determined by T, and is called the *spectral family of* T. If $f \in BV[\alpha,\beta]$ and $T = \int_{\mathbf{R}} \lambda dE_\lambda$ is well bounded then one defines $f(T) = \int_{\mathbf{R}} f(\lambda)dE_\lambda$. The map $f \mapsto f(T)$ is a continuous algebra homomorphism, called the *functional calculus associated with* T. It can be extended to the algebra $AC[\alpha,\beta]$ of absolutely continuous functions on $[\alpha,\beta]$ in the natural way. The following is an obvious generalization of [Do, Theorem 16.3(ii)].

Proposition 3.3. *Let $\{E_\lambda\}_{\lambda\in\mathbf{R}}$ be a spectral family of projections on X which is supported in $[\alpha,\beta]$. Let $f : [\alpha,\beta] \to \mathbf{R}$ be strictly monotone and continuous, and let $T = \int_{\mathbf{R}} f(\lambda)de_\lambda$. Let $S \in B(X)$. Then*

$$TS = ST \text{ if and only if } E_\lambda S = SE_\lambda \text{ for every } \lambda \in \mathbf{R}. \tag{3.17}$$

Let $\{p(\lambda)\}_{\lambda\in\mathbf{R}}, \{q(\lambda)\}_{\lambda\in\mathbf{R}}$ be two spectral measures of projections on the Hilbert space H, both supported on some finite interval in $\mathbf{R}_+$. Define a spectral measure $\mathcal{E} = \{E_\lambda\}_{\lambda\in\mathbf{R}}$ of projections on the Hilbert Schmidt class, $C_2(H)$, via

$$E_\lambda x = \int_{\mathbf{R}_+} p(\lambda t)x dq(t) + \int_{\mathbf{R}_+} dp(t)xq(\lambda t), \qquad 0 \leq \lambda < \infty, x \in C_2(H), \tag{3.18}$$

and $E_\lambda = 0$ for $\lambda < 0$. Recall ([GK2], [KP] and [A3]) that the *triangular projection*

$$(P_T x)(i,j) = \begin{cases} x(i,j); & i \leq j \\ 0; & i > j \end{cases}$$

is bounded on C_p for every $1 < p < \infty$. It follows from [A3,Th.4.11] that the projections E_λ are bounded on C_p, and moreover,

$$\|\mathcal{E}\|_{B(C_p)} := \sup_{\lambda \in \mathbf{R}} \|E_\lambda\|_{B(C_p)} \leq 2\|P_T\|_{B(C_p)} < \infty. \tag{3.19}$$

It follows easily that $\{E_\lambda\}_{\lambda \in \mathbf{R}}$ is a spectral family in $B(C_p)$ for every $1 < p < \infty$.

Proposition 3.4. *Let $f \in BV[0,1]$, let $\{E_\lambda\}_{\lambda \in \mathbf{R}}$ be as in (3.18) and let $x \in C_p, 1 < p < \infty$. Then*

$$\int_{[0,1]} f(\lambda) dE_\lambda x = \int\int_{\mathbf{R}_+ \times \mathbf{R}_+} f(\min\{\frac{s}{t}, \frac{t}{s}\}) dp(s) x \, dq(t). \tag{3.20}$$

Here "$\frac{0}{0}$"is interpreted as "0".

Proof: For every partition $0 = \lambda_0 < \lambda_1 < \cdots < \lambda_n = 1$,

$$\begin{aligned}
&\sum_{j=1}^{n} f(\lambda_j)(E_{\lambda_j} - E_{\lambda_{j-1}})x + f(0)E(0)x \\
&\quad = \int_{\mathbf{R}_+} \left[\sum_{j=1}^{n} f(\lambda_j)(p(\lambda_j t) - p(\lambda_{j-1} t))x + f(0)p(0)x \right] dq(t) \\
&\quad + \int_{\mathbf{R}_+} dp_s \left[x \sum_{j=1}^{n} f(\lambda_j)(q(\lambda_j s) - q(\lambda_{j-1} s)) + f(0)q(0) \right].
\end{aligned}$$

Thus, by the definition of the integral, a change of variable and Fubini's Theorem,

$$\begin{aligned}
\int_{[0,1]} f(\lambda) dE_\lambda x &= \int_{\mathbf{R}_+} \left(\int_0^t f\left(\frac{s}{t}\right) dp(s)x \right) dq(t) + \int_{\mathbf{R}_+} dp(s)x \int_0^s f\left(\frac{t}{s}\right) dq(t) \\
&= \int\int_{\mathbf{R}_+ \times \mathbf{R}_+} f\left(\min\left\{\frac{s}{t}, \frac{t}{s}\right\}\right) dp(s) x dq(t). \square
\end{aligned}$$

Suppose next that

$$\alpha_1 > \alpha_2 > \cdots > \alpha_j > \cdots > \alpha_0 = 0, \qquad \beta_1 > \beta_2 > \cdots > \beta_j > \cdots > \beta_0 = 0,$$

and $\{p_j\}_{j=0}^{\infty}$, $\{q_j\}_{j=0}^{\infty}$ are two sequences of orthogonal projections on H with $\sum_{j=0}^{\infty} p_j = I = \sum_{j=0}^{\infty} q_j$. For $0 \leq \lambda \leq 1$ define

$$p(\lambda) = \sum_{\alpha_j \leq \lambda} p_j, \qquad q(\lambda) = \sum_{\beta_j \leq \lambda} q_j.$$

Then $\{p(\lambda)\}_{\lambda\in\mathbf{R}}, \{q(\lambda)\}_{\lambda\in\mathbf{R}}$ are spectral measures on H, with support in a finite subinterval of $\mathbf{R}_+$. Let $\{E_\lambda\}_{\lambda\in\mathbf{R}}$ be defined by (3.18). Then

$$E_\lambda x = \sum_{j=0}^{\infty} \sum_{\substack{i=0 \\ \alpha_i \leq \lambda\beta_j}}^{\infty} p_i x q_j + \sum_{i=0}^{\infty} \sum_{\substack{j=0 \\ \beta_j \leq \lambda\alpha_i}}^{\infty} p_i x q_j, \tag{3.21}$$

for $0 \leq \lambda < \infty$ and $E_\lambda x = 0$ for $\lambda < 0$. By Proposition 3.4 we get in this context the following

Corollary 3.5. *In the special situation just described, for any $f \in BV[0,1]$ with $f(0) = 0$, and for every $x \in C_p, 1 < p < \infty$,*

$$\int_{[0,1]} f(\lambda) dE_\lambda = \sum_{0 \leq i,j < \infty} f(\min\{\frac{\alpha_i}{\beta_j}, \frac{\beta_j}{\alpha_i}\}) p_i\, x\, q_j. \tag{3.22}$$

Moreover $\int_{[0,1]} f(\lambda)dE_\lambda$ is bounded on C_p and

$$\| \int_{[0,1]} f(\lambda) dE_\lambda \|_{B(C_p)} \leq 2 \|P_T\|_{B(C_p)} \|f\|_{BV[0,1]}.$$

Theorem 3.6. *Let $0 < \tau, r, p_2 < \infty$ and let $1 < p_1 < \infty$ be so that $\frac{1}{p_2} = \frac{1}{p_1} + \frac{\tau}{r}$. Let a, b be positive elements in C_r with spectral decompositions $a = \sum_{j=1}^{\infty} \alpha_j p_j, b = \sum_{j=1}^{\infty} \beta_j q_j$, where $\alpha_1 > \alpha_2 > \cdots > \alpha_j > \cdots > 0, \beta_1 > \beta_2 > \cdots > \beta_j > \cdots > 0$. Set also $\alpha_0 = \beta_0 = 0$ and $p_0 = I - \sum_{j=1}^{\infty} p_j, q_0 = I - \sum_{j=1}^{\infty} q_j$. Define operators*

$$A_\tau^{\pm} x = \sum_{0 \leq i,j < \infty} \frac{\alpha_i^{1+\tau} \pm \beta_j^{1+\tau}}{\alpha_i \pm \beta_j} p_i x q_j. \tag{3.23}$$

Then $A_\tau^{\pm}$ are bounded operators from C_{p_1} into C_{p_2}, and

$$\|A_\tau^{\pm}\|_{B(C_{p_1}, C_{p_2})} \leq (6 + 4\tau) \|P_T\|^2_{B(C_{p_1})} (\|a\|_r^\tau + \|b\|_r^\tau). \tag{3.24}$$

Here if $\alpha_i = \beta_j$ for $0 < i, j$ then $\frac{\alpha_i^{1+\tau} - \beta_j^{1+\tau}}{\alpha_i - \beta_j}$ is interpreted as $(1+\tau)\alpha_i^\tau$, and the quotients corresponding to $i = j = 0$ are interpreted as 0.

Proof: Factorize $A^{\pm}_{\tau}$ as $A^{\pm}_{\tau} = S^{\pm}_{\tau} R^{\pm}_{\tau}$, where

$$\begin{aligned} S^{\pm}_{\tau} x &= \sum_{0 \le i,j < \infty} \max\{\alpha_i^{\tau}, \beta_j^{\tau}\} p_i\, x\, q_j \\ R^{\pm}_{\tau} x &= \sum_{0 \le i,j < \infty} f_{\pm}(\min\{\frac{\alpha_i}{\beta_j}, \frac{\beta_j}{\alpha_i}\}) p_i\, x\, q_j, \end{aligned}$$

and

$$f_{\pm}(t) = \frac{1 \pm t^{1+\tau}}{1 \pm t}.$$

Let

$$Tx = \sum_{\alpha_i \ge \beta_j} p_i x q_j.$$

Then T and $I - T$ are block triangular projections, and by [A3,Th.4.11],

$$\|T\|_{B(C_p)}, \|I - T\|_{B(C_{p_1})} \le \|P_T\|_{B(C_{p_1})}.$$

Since

$$S^{\pm}_{\tau} x = a^{\tau}(Tx) + ((I - T)x) b^{\tau},$$

we get by the Hölder inequality and $\dfrac{1}{p_2} = \dfrac{1}{p_1} + \dfrac{\tau}{r}$,

$$\begin{aligned} \|S^{\pm}_{\tau} x\|_{p_2} &\le \|a\|_r^{\tau} \|Tx\|_{p_1} + \|(I - T)x\|_{p_1} \|b\|_r^{\tau} \\ &\le \|P_T\|_{B(C_1)} (\|a\|_r^{\tau} + \|b\|_r^{\tau}) \|x\|_{p_1}. \end{aligned}$$

Thus $\|S^{\pm}_{\tau}\|_{B(C_{p_1}, C_{p_2})} \le \|P_T\|_{B(C_1)}(\|a\|_r^{\tau} + \|b\|_r^{\tau})$. Next, it is easy to verify that $\|f_+\|_{BV[0,1]} \le 3$ and $\|f_-\|_{BV[0,1]} \le 1 + 2\tau$. Hence, by Corollary 3.5, (3.16) and (3.19) we see that

$$\|R^{\pm}_{\tau}\|_{B(C_{p_1})} \le 2\|P_T\|_{B(C_{p_1})} \max(3, 1 + 2\tau).$$

Finally, since $A^{\pm}_{\tau} = S^{\pm}_{\tau} R^{\pm}_{\tau}$,

$$\begin{aligned} \|A^{\pm}_{\tau}\|_{B(C_{p_1}, C_{p_2})} &\le \|S^{\pm}_{\tau}\|_{B(C_{p_1}, C_{p_2})} \|R^{\pm}_{\tau}\|_{B(C_{p_1})} \\ &\le 2 \max\{3, 1 + 2\tau\} \|P_T\|^2_{B(C_{p_1})} (\|a\|_r^{\tau} + \|b\|_r^{\tau}). \square \end{aligned}$$

Remark. A result analogous to (3.23) with A_τ^- is obtained in [A2] for a larger class of functions. The techniques of [A2] are different and depend on factorization through the Hilbert space C_2.

Corollary 3.7. *Let τ, r, p_1, p_2 be as in Theorem 3.6. Let $x \in C_r$ have a Schmidt series $x = \sum_{j=1}^\infty \alpha_j v_j$. Let $\{P_{i,j}\}_{0 \le i \le j < \infty}$ be the joint Peirce projections associated with $\{v_j\}$ and let*

$$M_{1+\tau,x}^{\pm} := \sum_{0 \le i \le j < \infty} \frac{\alpha_i^{1+\tau} \pm \alpha_j^{1+\tau}}{\alpha_i \pm \alpha_j} P_{i,j}. \tag{3.25}$$

Then $M_{1+\tau,x}^{\pm}$ are bounded operators fron C_{p_1} into C_{p_2}, and

$$\|M_{1+\tau,x}^{\pm}\|_{B(C_{p_1},C_{p_2})} \le (12+8\tau)\|P_T\|_{B(C_{P_1})}^2 \|x\|_r^\tau. \tag{3.26}$$

In particular, if $2 < p < \infty$ and $r = p_1 = p, \tau = p-2$ and $p_2 = q \quad (\frac{1}{p} + \frac{1}{q} = 1)$, then the operators

$$M_{p-1,x}^{\pm} := \sum_{0 \le i \le j < \infty} \frac{\alpha_i^{p-1} \pm \alpha_j^{p-1}}{\alpha_i \pm \alpha_j} P_{i,j} \tag{3.27}$$

are bounded from C_p into C_q and

$$\|M_{p-1,x}^{\pm}\|_{B(C_p,C_q)} \le (4p+8)\|P_T\|_{B(C_p)}^2 \|x\|_p^{p-2}. \tag{3.28}$$

Proof: Use Theorem 3.6 with $a = |x^*|, b = |x|$. Thus $\alpha_j = \beta_j$ for all j. Notice that $P_{i,j}y = p_i y q_j + p_j y q_i$ for $1 \le i < j$. Hence $M_{1+\tau,x}^{\pm}$ are simply the $A_\tau^{\pm}$ of Theorem 3.5. This yields (3.26). The rest is obvious. □

Next, we need a perturbation formula from [ABF]. Let $x, y \in C_\infty$ have Schmidt series

$$x = \sum_j \alpha_j v_j \ , \ y = \sum_j \beta_j w_j$$

where $\{\alpha_j\}, \{\beta_j\}$ are decreasing sequences of positive numbers and $\{v_j\}, \{w_j\}$ sequences of orthogonal, finite-rank tripotents. Let $v = \sum_j v_j$ and $w = \sum_j w_j$ and define *joint symmetrization and anti-symmetrization operators* $S_{v,w}$ and $A_{v,w}$ by

$$S_{v,w}(z) = \frac{1}{2}(z + wz^*v) \ , \ A_{v,w}(z) = \frac{1}{2}(z - wz^*v). \tag{3.29}$$

Let f be an odd function in $C^{1+\epsilon}(\mathbf{R})$ for some $\epsilon > 0$. Define $m_f^{\pm}(s,t) = (f(s) \pm f(t))/(s \pm t)$, with the usual convention in case the denominator vanishes. Consider the *joint Peirce multipliers*

$$M_{f,x,y}^{\pm} z = \sum_{i,j} m_f^{\pm}(\beta_i, \alpha_j) w_i w_i^* z v_j^* v_j. \tag{3.30}$$

The *perturbation formula* [ABF,Th.3.2] that we need is

$$f(y) - f(x) = M_{f,x,y}^{-} S_{v,w}(y-x) + M_{f,x,y}^{+} A_{v,w}(y-x). \tag{3.31}$$

Proposition 3.8. *Let $0 < \tau_1, p_2 < \infty$ and let $1 < p_1 < \infty$ be so that $\frac{1}{p_2} = \frac{1+\tau}{p_1}$. Let $f(\lambda) = sgn(\lambda)|\lambda|^{1+\tau}, \lambda \in \mathbf{R}$. Let $x, y \in C_{p_1}$. Then for $0 < |t| \le 1$,*

(i) $\|\frac{f(x+ty)-f(x)}{t}\|_{p_2} \le \gamma(\|x\|_{p_1} + \|y\|_{p_1})^{\tau} \|P_T\|^2_{B(C_{p_1})} \|y\|_{p_1}$, *with γ depending only on τ;*

(ii) *If, moreover $1 < p_2 < \infty$, then the derivative $\frac{d}{dt} f(x+ty)_{|t=0}$ exists in the weak topology of C_{p_2}, and*

$$\frac{d}{dt} f(x+ty)_{|t=0} = (M_{f,x}^{-} S_v + M_{f,x}^{+} A_v) y \tag{3.32}$$

where $M_{f,x}^{\pm} = M_{f,x,x}^{\pm}, S_v = S_{v,v^*}$ and $A_v = A_{v,v^*}$ are the operators defined by (3.12).

Proof: Let $x = v|x|$ and $x + ty = w|x+ty|$ be the polar decompositions of x and $x + ty$. By the perturbation formula (3.31) and Theorem 3.5,

$$\begin{aligned} \|\frac{f(x+ty)-f(x)}{t}\|_{p_2} &= \|M_{f,x+ty}^{-} S_{v,w} y + M_{f,x,x+ty}^{+} A_{v,w} y\|_{p_2} \\ &\le \gamma(\|x\|_{p_1} + |t|\|y\|_{p_1})^{\tau} \|P_T\|^2_{B(C_{p_1})} \|y\|_{p_1} \end{aligned}$$

where $\gamma = 8\max\{3, 1+2\tau\}$. This proves (i). Assume now that $1 < p_2 < \infty$, then $1 + \tau < p_1$ and $C_{p_2}^* = C_{p_1/(p_1-1-\tau)}$. Thus $C_{p_1}^* \subset C_{p_2}^*$ and $C_{p_1}^*$ is dense in $C_{p_2}^*$. By Theorem 3.2 and the perturbation formula (3.31),

$$\frac{d}{dt} f(x+ty)_{|t=0} = (M_{f,x}^{-} S_v + M_{f,x}^{+} A_v) y$$

where the limit in the differentiation process is taken in the norm of C_{p_1}. Thus, for every $z \in C^*_{p_1}$

$$\lim_{t\to 0}\langle \frac{f(x+ty)-f(x)}{t}, z\rangle = \langle (M^-_{f,x}S_v + M^+_{f,x}A_v)y, z\rangle.$$

By part (i) and the fact that $C^*_{p_1}$ is dense in $C^*_{p_2}$, we see that the last limit and equality hold also for $z \in C^*_{p_2}$. □

Corollary 3.9. *Let $2 < p < \infty$ and let $f(\lambda) = sgn(\lambda)|\lambda|^{p-1}$. Then the triple map $f : C_p \to C_q$ $(\frac{1}{p} + \frac{1}{q} = 1), f(x) = x^{p-1}$, is Gateaux differentiable in the weak topology of C_q, and its derivative is given by (3.32). Consequently the map N_p : $C_p \to C_q$ is Gateaux differentiable in the weak topology of C_q. Moreover, if $X \subseteq C_p$ is an N-convex subspace then for every $x, y \in X$, $N'_p(x)y := \frac{d}{dt}N_p(x+ty)_{|t=0} \in N_p(X)$.*

Proof: Apply Proposition 3.8 with $p_1 = p, p_2 = q$ and $\tau = p-2$. Then use the fact that $N_p(X)$ is a closed linear subspace of C_q and that every norm-closed subspace is also weakly closed. □

Remark. It is conjectured that the differentiability properties of the norm of C_p are exactly the same as those of L^p, (see [BF] for the L^p-result). This is stated without a proof in [T2]. In particular, the weak differentiability in Corollary 3.9 is actually a differentiability in the norm of C_q.

We will need one more result concerning the *second derivative* of the map $x \mapsto x^{2+\tau}, 0 < \tau$. To give perspective we prefer to describe the result concerning the n-th order derivative for arbitrary n. Let J be a finite or infinite interval and let $C^n(J)$ denote the space of n-th order continuously differentiable function on J. For $0 < \epsilon < 1, C^{n+\epsilon}(J)$ denotes the space of all $f \in C^n(T)$ for which $f^{(n)} \in Lip_\epsilon(J)$. For $f \in C^n(J)$ the *higher order divided differences* $f^{[k]}(\lambda_0, \lambda_1, \cdots, \lambda_k), 1 \le k \le n,, \lambda_j \in$

J, are defined inductively by

$$f^{[k]}(\lambda_0, \lambda_1, \cdots, \lambda_k) = \begin{cases} \dfrac{f^{[k-1]}(\lambda_0, \cdots, \lambda_{k-2}, \lambda_{k-1}) - f^{[k-1]}(\lambda_0, \cdots, \lambda_{k-2}, \lambda_k)}{\lambda_{k-1} - \lambda_k}; & \lambda_{k-1} \neq \lambda_k \\ \dfrac{\partial}{\partial t} f^{[k-1]}(\lambda_0, \cdots, \lambda_{k-2}, t)_{|t=\lambda_k}; & \lambda_{k-1} = \lambda_k. \end{cases}$$

Theorem 3.10. *Let n be a positive integer and let $0 < \epsilon < 1$. Let $f \in C^{n+\epsilon}(J)$ and let $b_1, b_2, \cdots, b_n, a \in B(H)_{s.a.}$ with $\sigma(a) \subset J$ and $a = \int_{\mathbf{R}} \lambda de_\lambda$.*

(i) *The integral*

$$F(a; b_1, b_2, \cdots, b_n) := \int_{\mathbf{R}^{n+1}} f^{[n]}(\lambda_0, \lambda_1, \cdots, \lambda_n) de_{\lambda_0} b_1 de_{\lambda_1} b_2 \cdots b_n de_{\lambda_n} \quad (3.34)$$

converges in the operator norm and

$$\|F(a; b_1, \cdots, b_n)\| \leq C(J, n, \epsilon) \|f\|_{C^{n+\epsilon}} \|b_1\| \|b_2\| \cdots \|b_n\|. \quad (3.35)$$

(ii) *The function $J^n \ni (t_1, \cdots, t_n) \mapsto f(a + \sum_{j=1}^n t_j b_j)$ is n times continuously differentiable in the operator norm and*

$$\frac{\partial^n}{\partial t_1 \partial t_2 \cdots \partial t_n} f(a + \sum_{j=1}^n t_j b_j)_{|t_1=t_2=\cdots=t_n=0} = \sum_\pi F(a; b_{\pi(1)}, b_{\pi(2)}, \cdots, b_{\pi(n)}) \quad (3.36)$$

where the sum ranges over all permutations π of $\{1, 2, \cdots, n\}$.

(iii) *If $b_j \in (C_{p_j})_{s.a}, j = 1, \cdots, n$, and $\frac{1}{r} = \min\{1; \frac{1}{p_1} + \frac{1}{p_2} + \cdots + \frac{1}{p_j}\}$, then the integral (3.34) converges in C_r,*

$$\|F(a; b_1, b_2, \cdots, b_n)\|_r \leq C(J, n, \epsilon) \|f\|_{C^{n+\epsilon}} \|b_1\|_{p_1} \cdot \|b_2\|_{p_2} \cdots \|b_n\|_{p_n} \quad (3.37)$$

and the differentiation in (3.5) is in the norm of C_r;

(iv) *For $b \in B(H)_{s.a.}$,*

$$\frac{1}{n!} \frac{d^n}{dt^n} f(a + tb)_{|t=0} = F(a; b, b, \cdots, b). \quad (3.38)$$

Parts (i), (ii) and (iv) of Theorem 3.10 are proved in [DK] under stronger assumptions on f. The theory of the multiple Stieltjes operator integrals as in (3.34) is developed in [SS] and [Pa]. This theory is combined in [SS] with the methods of [BSIII] to yield Theorem 3.10.

If $a = \sum_j \lambda_j e_j$ is the spectral decomposition of $a \in B(H)_{s.a.}$, with distinct eigenvalues $\{\lambda_j\}$ and corresponding spectral projections $\{e_j\}$, then (3.38) takes the form

$$\frac{1}{n!}\frac{d^n}{dt^n} f(a+tb)_{|t=0} = \sum_{j_0,j_1,\cdots,j_n} f^{[n]}(\lambda_{j_0}, \lambda_{j_1}, \cdots, \lambda_{j_n}) e_{j_0} b e_{j_1} b \cdots b e_{j_n}. \tag{3.39}$$

4. The Connection between a Contractive Projection on C_p and the Peirce Projections Associated with the Elements in its Range

Let P be a contractive projection in $C_p, 1 < p < \infty$, $p \neq 2$, and let P^* be the dual projection in C_q, $q^{-1} + p^{-1} = 1$. Let $R(P)$ and $R(P^*)$ denote the ranges of P and P^* respectively. We fix $x \in R(P)$ with $\|x\|_p = 1$ and a Schmidt series

$$x = \sum_j \alpha_j v_j \tag{4.1}$$

where $\alpha_1 > \alpha_2 > \ldots > \alpha_j > \ldots..$ are the non-zero singular numbers of x and $\{v_j\}$ are orthogonal, finite rank tripotents. We denote $\alpha_0 = 0$ for convenience . As in (2.14), let $\{P_{i,j}\}_{0 \leq i \leq j}$ be the joint Peirce projections associated with the $\{v_j\}$. Let $x = v|x|$ be the polar decomposition of x, where $v = v(x) = \sum_j v_j$, and let $P_k = P_k(v), k = 0, 1, 2$, be the Peirce projections of v. Notice that

$$P_2 = \sum_{1 \leq i \leq j} P_{i,j} \ , \quad P_1 = \sum_{1 \leq j} P_{0,j} \ , \quad P_0 = P_{0,0}. \tag{4.2}$$

As usual, $Q = Q(v)$ is the operator $Qy = vy^*v = \{vyv\}$.

For $0 \leq \lambda \leq 1$ we let

$$A_\lambda = \{(i,j); 0 \leq i \leq j, 0 < j, \alpha_j \alpha_i^{-1} = \lambda\} \tag{4.3}$$

and

$$L_\lambda = \sum_{(i,j)\in A_\lambda} P_{i,j}. \tag{4.4}$$

The main result of this section is the following theorem .

Theorem 4.1 *Let $1 < p < \infty$, $p \neq 2$, and let $x \in R(P)$ be as above. Then*

(i) $PL_\lambda = L_\lambda P \quad ,0 \leq \lambda \leq 1;$

(ii) $PP_k = P_k P \quad ,k = 0,1,2;$

(iii) $PQ = QP.$

Proof: Since the family of the projections $\{L_\lambda\}_{0\leq\lambda\leq 1}$ associated with x is the same as the family associated with $N_p(x)$, we assume without loss of generality that $2 < p < \infty$. Let $S := (I+Q)/2$ and $A := (I-Q)/2$ be the symmetrization and anti-symmetrization operators associated with Q. Define operators

$$M_- = \sum_{0\leq i\leq j} \frac{\alpha_i^{p-1} - \alpha_j^{p-1}}{\alpha_i - \alpha_j} P_{i,j} \tag{4.5}$$

and

$$M_+ = \sum_{0\leq i\leq j} \frac{\alpha_i^{p-1} + \alpha_j^{p-1}}{\alpha_i + \alpha_j} P_{i,j} \tag{4.6}$$

where $(\alpha_i^{p-1} - \alpha_j^{p-1})/(\alpha_i - \alpha_j)$ means $(p-1)\alpha_i^{p-2}$ if $i = j > 0$ and 0 if $i = j = 0$. By Corollary 3.7, M_+ and M_- map C_p continously into C_q.

Let $y \in R(P)$. By Corollary 3.9,

$$\frac{d}{dt}(x+ty)^{p-1}_{|t=0} = M_- Sy + M_+ Ay \in R(P^*). \tag{4.7}$$

Since Q is conjugate-linear, $S(iy) = iAy$ and $A(iy) = iSy$. Using the fact that $R(P)$ is a complex-linear subspace (so, $iy \in R(P)$ as well) we get by the same arguments

$$\frac{1}{i}\frac{d}{dt}(x+ity)^{p-1}_{|t=0} = M_- Ay + M_+ Sy \in R(P^*). \tag{4.8}$$

Define

$$M_1 = \sum_{0\leq i\leq j} \frac{\alpha_i^p - \alpha_j^p}{\alpha_i^2 - \alpha_j^2} P_{i,j} \ , \tag{4.9}$$

and

$$M_2 = \sum_{0 \le i \le j} \alpha_i \alpha_j \frac{\alpha_i^{p-2} - \alpha_j^{p-2}}{\alpha_i^2 - \alpha_j^2} P_{i,j} \ , \tag{4.10}$$

where the quotients are defined by the appropriate limits when $i = j$. Notice that M_1, M_2 are self-adjoint with respect to the Hilbert-Schmidt inner product $\langle a, b \rangle = \text{trace}(ab^*)$, i.e.

$$\langle M_\nu a, b \rangle = \langle a, M_\nu b \rangle \ , a, b \in C_p, \nu = 1, 2. \tag{4.11}$$

Notice also that $M_1 = M_- + M_+$ and $M_2 = M_- - M_+$. Since $S + A = I$ and $S - A = Q$, (4.7) and (4.8) imply that M_1 and $M_2 Q$ map $R(P)$ into $R(P^*)$, i.e.,

$$M_1 P = P^* M_1 P \quad , \quad M_2 Q P = P^* M_2 Q P. \tag{4.12}$$

Define continuous sesquilinear forms $\langle \cdot, \cdot \rangle_\nu$, $(\nu = 1, 2)$ in C_p by

$$\langle a, b \rangle_\nu = \langle M_\nu a, b \rangle = tr((M_\nu a) b^*); \ a, b \in C_p. \tag{4.13}$$

We claim that $(\cdot, \cdot)_\nu$ are positive. Indeed if $a \in C_p$ has a Peirce expansion $a = \sum_{0 \le i \le j} a_{i,j}$ where $a_{i,j} = P_{i,j} a$, then

$$(a, a)_1 = \sum_{0 \le i \le j, 0 < j} \frac{\alpha_i^p - \alpha_j^p}{\alpha_i^2 - \alpha_j^2} \|a_{i,j}\|_2^2 \quad \ge 0. \tag{4.14}$$

Similarly $(a, a)_2 \ge 0$. Notice that $(a, a)_1 = 0$ if and only if $a \in R(P_0)$, that is $(\cdot, \cdot)_1$ is faithful on $R(P_1 + P_2)$. Similarly, $(\cdot, \cdot)_2$ is faithful on $R(P_2)$.

Lemma 4.2 *(i) P is bounded with respect to $(\cdot, \cdot)_1$ and $(\cdot, \cdot)_2$;*
(ii) P is self adjoint with respect to $(\cdot, \cdot)_1$, i.e. for all $a, b \in C_p$,

$$(Pa, b)_1 = (a, Pb)_1 \quad ; \tag{4.15}$$

(iii) For all $a, b \in C_p$,

$$(Pa, b)_2 = (a, \text{QPQb})_2 \ . \tag{4.16}$$

Proof: The first statement follows from the fact that $M_1, M_2 \in B(C_p, C_q)$. Let $a, b \in C_p$. By (4.11) and the fact that M_1 is self-adjoint we have

$$\langle Pa, b \rangle_1 = \langle M_1 Pa, b \rangle = \langle P^* M_1 Pa, b \rangle = \langle M_1 Pa, Pb \rangle = \langle Pa, M_1 Pb \rangle$$

$$= \langle a, P^* M_1 P b\rangle = \langle a, M_1 P b\rangle = \langle M_1 a, P b\rangle = \langle a, P b\rangle_1.$$

Similarly, since $QM_2 = M_2 Q$ and $M_2 = M_2 P_2 = M_2 Q^2$, we get by (4.12) and the fact that $\langle Qz, y\rangle = \langle Qy, z\rangle$ for all $y, z \in C_p$,

$$\begin{aligned} \langle Pa, b\rangle_2 &= \langle M_2 Pa, b\rangle = \langle QM_2 QPa, b\rangle = \langle QP^* M_2 QPa, b\rangle \\ &= \langle Qb, P^* M_2 QPa\rangle = \langle PQb, M_2 QPa\rangle = \langle M_2 Pa, QPQb\rangle \\ &= \langle a, M_2 QPQb\rangle = \langle a, QPQb\rangle_2. \ \square \end{aligned}$$

Define an operator M by

$$M = \sum_{1 \leq i \leq j} \alpha_i \alpha_j \frac{\alpha_i^{p-2} - \alpha_j^{p-2}}{\alpha_i^p - \alpha_j^p} P_{i,j}. \tag{4.17}$$

Clearly, $M_1 M = M_2$. Let us define a function φ by $\varphi(0) = (p-2)/p$ and

$$\varphi(t) = t(1 - t^{p-2})/(1 - t^p), \ 0 < t \leq 1. \tag{4.18}$$

Then φ is a strictly increasing, continously differentiable function on [0,1], and

$$M = \sum_{1 \leq i \leq j} \varphi(\alpha_j \alpha_i^{-1}) P_{i,j}. \tag{4.19}$$

By Corollary 3.5, M is a bounded operator on C_p . Obviously,

$$(a, b)_2 = (Ma, b)_1, \quad a, b \in C_p. \tag{4.20}$$

We claim that

$$(PMa, b)_1 = (MQPQa, b)_1 \ , \quad a, b \in C_p. \tag{4.21}$$

Indeed, by Lemma 4.2 and (4,20) we have

$$(PMa, b)_1 = (Ma, Pb)_1 = (a, Pb)_2 = (QPQa, b)_2 = (MQPQa, b)_1.$$

Since $(\cdot, \cdot)_1$ is faithful on $R(P_1 + P_2)$, we get from (4.21)

$$(P_1 + P_2) PM = MQPQ. \tag{4.22}$$

Thus,

$$P_1 P M = 0 \tag{4.23}$$

and

$$P_2 P M = MQPQ. \tag{4.24}$$

Let

$$\Lambda = \{\alpha_j \alpha_i^{-1};\, 1 \leq i \leq j\}. \tag{4.25}$$

Then Λ consists of the non-zero spectral points of M, all of which are eigenvalues. The corresponding eigenprojections are $\{L_\lambda\,;\,\lambda \in \Lambda\}$. Since $P_2 = \sum_{\lambda \in \Lambda} L_\lambda$, we get from (4.23)

$$P_1 P P_2 = 0. \tag{4.26}$$

Using the fact that P_1, P_2 and P are self-adjoint with respect to $(\cdot,\cdot)_1$, and that $(\cdot,\cdot)_1$ is faithful on $R(P_1 + P_2)$, (4.26) implies

$$P_2 P P_1 = 0. \tag{4.27}$$

Also, the self-adjointness of P with respect to $(\cdot,\cdot)_1$ implies that P maps the kernel of $(\cdot,\cdot)_1$ into itself, i.e. $P(R(P_0)) \subseteq R(P_0)$. Thus, $PP_0 = P_0 P P_0$. Corollary 1.7 now implies that

$$PP_0 = P_0 P. \tag{4.28}$$

It follows from (4.26),(4.27) and(4.28) that

$$PP_1 = P_1 P \quad \text{and} \quad PP_2 = P_2 P. \tag{4.29}$$

Notice that this completes the proof of part (ii) of Theorem 4.1.

We claim now that

$$PM^2 = M^2 P. \tag{4.30}$$

Indeed, (4.24) and (4.29) imply

$$PM = MQPQ. \tag{4.31}$$

Since M commutes with Q, we get by using (4.31)

$$PM^2 = MQPQM = MQPMQ = MQMQP = M^2P.$$

Next, if $E_\lambda = \sum_{\tau \leq \lambda} L_\tau$ for $0 \leq \lambda \leq 1$, $E_\lambda = 0$ for $\lambda < 0$ and $E_\lambda = I$ for $1 < \lambda$ then

$$M^2 = \sum_{\lambda \in \Lambda} \varphi(\lambda) L_\lambda = \int_{\mathbf{R}} \varphi(\lambda) dE_\lambda. \tag{4.32}$$

Thus, Proposition 3.3 and the fact that φ is strictly increasing imply

$$PE_\lambda = E_\lambda P, \; PL_\lambda = L_\lambda P \quad , \; \lambda \in \Lambda, \tag{4.33}$$

and

$$P\psi(M^2) = \psi(M^2)P \tag{4.34}$$

for every function of bounded variation ψ on $[0, (p-2)/p]$. In particular,

$$PM = MP. \tag{4.35}$$

Notice that $L_0 = P_1$. So (4.29) and (4.33) prove part (i) in Theorem 4.1. Using (4.31) and (4.35) we get

$$MPQ = MQP. \tag{4.36}$$

Since M is one-to-one on $R(P_2)$, it follows that

$$PQ = QP. \tag{4.37}$$

This concludes the proof of Theorem 4.1. □

Remark: *(i)* The formula $PQ = QP$ proved in Therem 4.1(*iii*) means that P is self-adjoint with respect to the conjugation $Q = Q(v)$ corresponding to the element $x \in R(P)$. That is, P *is locally self-adjoint.*

(*ii*) One can introduce an ordering on $R(P_2(x))$ by requiring that an element $y \in R(P_2(x))$ is positive if $v(x)^* y$ is positive in the ordinary sense. It follows aposteriori from the Main Theorem that P *is positive with respect to the partial* *ordering associated with each element $x \in R(P)$, that is, P is locally positive.* The continuation of the proof of the Main Theorem (sections 5, 6 and 7) uses in a crucial way the atomic nature of C_p. It is interesting to give a direct and conceptual proof of the the local positivity without using the atomic nature of C_p.

5. Existence of atoms

Let X be a subspace of C_p $(1 < p < \infty, p \neq 2)$, and let x be a norm one element in X. Denote by $P_k(x) = P_k(v(x))$ the Peirce projections of x (or, of $v(x)$, the partial isometry in the polar decomposition of x) and let $X_k(x) = \{y \in X : P_k(x)y = y\}, k = 0,1,2$. Let $x = \sum_{j=1}^{n} \alpha_i v_j$ $(n \leq \infty)$ be the Schmidt series of x, where $\alpha_1 > \alpha_2 > \ldots > \alpha_j > \ldots > 0$ and $\{v_j\}_{j=1}^{n}$ are orthogonal finite rank tripotents (=partial isometries). Define

$$Dy = \sum_{j=1}^{n} P_2(v_j)y = \sum_{j=1}^{n} v_j v_j^* y v_j^* v_j. \tag{5.1}$$

D is the *(block) diagonal projection* associated with the family of tripotents $\{v_j\}$.

Definition 5.1 Let X and x be as above. Then

(i) x is said to be an *atom* if $X_2(x)$ is one-dimensional, i.e. $y \in X$ and $P_2(x)y = y$ imply that $y = \lambda x$ for some $\lambda \in \mathbf{C}$;

(ii) x is said to be an *atom in its diagonal* if $y \in X$ and $Dy = y$ imply that $y = \lambda x$ for some $\lambda \in \mathbf{C}$;

(iii) x is said to have *minimal support* if $0 \neq y \in X_2(x)$ implies $P_2(y) = P_2(x)$;

(iv) x is said to be *indecomposable* if it cannot be written as the sum of two non-zero orthogonal elements of X.

Of course, the four properties in Definition 5.1 are relative to X. If $X = R(P)$, where P is a contractive projection on C_p, then an atom of X is considered also as an atom of P. Also, x is an atom in its diagonal if and only if it is an atom of $D \wedge P$, the infimum of D and P.

While the four properties in the Definition 5.1 are not equivalent if X is a general subspace, they are equivalent for $X = R(P)$. The following theorem is the main results in this section.

Theorem 5.2 *Let P be a contractive projection on C_p $(1 < p < \infty, p \neq 2)$ and let $X = R(P)$. Then any normalized element $x \in X$ which satisfies one of the four properties (i)-(iv) of Definition 5.1, satisfies all the rest. Thus properties (i)-(iv) of Definition 5.1 are equivalent in N-convex spaces.*

Theorem 5.2 will be used to establish the following result

Theorem 5.3 *Every contractive projection P on C_p $(1 < p < \infty, p \neq 2)$ has an atom.*

Some parts of the proof of Theorem 5.2 are elementary. The implications $(i) \Rightarrow (ii), (i) \Rightarrow (iii)$ and $(ii) \Rightarrow (iv)$ are straightforward. To prove $(iii) \Rightarrow (ii)$, let $y \in X$ be so that $Dy = y$. Let $y_j = P_2(v_j)y, 1 \leq j \leq n$. If $y = 0$ we are done. Otherwise, we must have $P_2(y_j) = P_2(v_j)$ for all j, since x is assumed to have minimal support. Since v_1 and y_1 have finite rank, there is a $\lambda \in \mathbf{C}$ so that rank $(y_1 - \lambda\alpha_1 v_1) <$ rank (v_1). It follows that $P_2(y - \lambda x) < P_2(x)$. Since x has a minimal support, this implies that $y - \lambda x = 0$, completing the proof.

The proof of $(iv) \Rightarrow (ii)$ is also elementary. Let $x = \sum_{j=1}^{n} \alpha_j v_j$ $(n \leq \infty)$ be the spectral decomposition of x, and let $0 \neq y \in X$ be so that $Dy = y$. We must show that $y = \lambda x$ for some $\lambda \in \mathbf{C}$. We may assume that $x \geq 0$ and $y = y^*$, since $Q(\, v(x)\,)$ commutes with P. Write $y = \sum_{j=1}^{n} y_j$ where $y_j = P_2(v_j)y$. Thus $y_j = y_j^*$ for all j. If y is not proportional to x we can replace y be an appropriate linear combination, say z of x and y which has a non-trivial kernel. It follows that $x = P_2(z)x + P_0(z)x$

is a non-trivial decomposition of x into the sum of two orthogonal elements of X. This contradicts the fact that x is indecomposable in X and completes the proof of $(iv) \Rightarrow (ii)$.

In order to complete the proof of Theorem 5.2, it remains to prove the implication $(ii) \Rightarrow (i)$.

Lemma 5.4 *Let $x \in X$ be an atom in its diagonal. Then x is an atom of X.*

We pospone the proof of Lemma 5.4 and show first how Theorem 5.3 follows from Theorem 5.2 and the following result. Here the notion of indecomposable elements is extended in the obvious way to any subspace of C_p.

Lemma 5.5 *Let Y be a non-zero, closed subspace of $C_p, 1 \leq p \leq \infty$. Then any $0 \neq y \in Y$ can be written as the sum $y = \sum_{j\in J} y_j$, where $\{y_j\}_{j\in J}$ are pairwise orthogonal indecomposable elements of Y. In particular, Y has an indecomposable element.*

Proof: Let $y = \sum_{i=1}^{n} \beta_i u_i$ be the Schmidt series of y, with $\beta_1 > \beta_2 > \ldots$, $\{u_i\}$ orthogonal tripotents of finite rank, and $n \leq \infty$. Consider the set A of all sequences $\{y_j\}_{j\in J}$ of pairwise orthogonal, non-zero elements of Y so that $y = \sum_{j\in J} y_j$. We order A by refinement, that is $\{y_j\}_{j\in J} \leq \{z_k\}_{k\in K}$ if $y_j = \sum_{k\in K_j} z_k$ for every j (of course, the set $\{K_j\}$ forms a partition of K). If $\{y_j\}_{j\in J}$ is a maximal element of A, then each y_j is indecomposable, so $y = \sum_{j\in J} y_j$ is the desired decomposition of y. Our goal is to prove the existence of a maximal element in A via Zorn's lemma.

Notice that if $\{y_j\}_{j\in J} \in A$ then $y_j = \sum_i \beta_i u_{ij}$, where $\{u_{ij}; j \in J, i = 1, 2 \ldots\}$ are pairwise orthogonal tripotents with $u_i = \sum_{j\in J} v_{ij}$.

To prove the existence of a maximal element in A we shall show that every linearly ordered subset of A has an upper bound in A. Notice that if $\{y_j\}_{j\in J}$ and $\{z_k\}_{k\in K}$ are in $A, \{y_j\}_{j\in J} \leq \{z_k\}_{k\in K}$ and $y_j = \sum_{i=1}^{n} \beta_i u_{i,j}$, $z_k = \sum_{i=1}^{n} \beta_i w_{i,k}$, then $y_j = \sum_{k\in K_j} z_k$ implies $u_{i,j} = \sum_{k\in K_j} w_{i,k}$.

Thus the refinement of the partition of y implies the refinement of the partition of each v_i. It is crucial to observe that the v_i have finite rank, and so the partitions of each v_i must stabilize after finitely many steps. It is now clear how to define the upper bound in A of any linearly ordered subset. Application of Zorn's lemma completes the proof. □

Remark: *Lemma 5.5 holds in every separable unitary ideal and in particular in C_∞; the proof is the same. Notice that Lemmas 5.5, 5.4 and Theorem 2.2 imply Theorem 5.3.*

In the proof of Lemma 5.4 it will be convenient to assume that $2 < p < \infty$. The following result shows that there is no loss of generality in assuming this.

Proposition 5.6: *Let $x \in X = R(P), \|x\| = 1$. Then x enjoys one of the properties (i)-(iv) of Definition 5.1 if and only if the same is true for $N_p(x)$ with respect to $R(P^*)$.*

The easy proof is omitted.

Proof of Lemma 5.4

Let x be a normalized element of X which is an atom in its diagonal. We shall prove that x is an atom of X. Without loss of generality we assume that $2 < p < \infty$. Clearly, x is an atom of P if and only if it is an atom of $P_2(x)P$, thus we assume without loss of generality that $P = P_2(x)P$ and so $X = X_2(x)$. We also assume for convenience that $x \geq 0$; the general case can be reduced to this case by replacing P by the projection $y \mapsto v(x)^*P(v(x)y)$.

Let $x = \sum_j \alpha_j p_j$ be the spectral decomposition of x, where $\alpha_1 > \alpha_2 > \ldots$, $\{p_j\}$ are orthogonal projections of finite rank and $n \leq \infty$. Let $\{P_{i,j}\}_{0 \leq i \leq j < n}$ be the joint Peirce projections associated with the family $\{p_j\}$. For each $o < \tau \leq 1$, consider

$$A_\tau = \{(i,j);\ 1 \leq i \leq j < n, \quad \frac{\alpha_j}{\alpha_i} = \tau\} \tag{5.2}$$

and define

$$L_\tau y = \sum_{(i,j)\in A_\tau} P_{i,j} y \quad 0 < \tau \le 1. \tag{5.3}$$

Notice that $E_\lambda := \sum_{\tau\le\lambda} L_\tau$ are a special case of the projections defined by (3.21).

Let $y \in X$, i.e. $y = Py = P_2(x)y$. By Theorem 4.1, $Q(x)y = v(x)y^*v(x) \in X$. But $x \ge 0$ and so $Q(x)y = y^*$. Thus $y = y_1 + iy_2$ with y_1, y_2 selfadjoint elements of X. It will be enough to show that every selfadjoint element of X is proportional to x. We assume therefore that $y = y^*$.

By Theorem 4.1, $y_\tau := L_\tau y \in X$. Since x is an atom in its diagonal and $A_1 = \{(i,i); 1 \le i \le n\}$ we see that $y_1 = \lambda x$ for some $\lambda \in \mathbf{R}$. Since $y = \sum_{0<\tau\le 1} y_\tau$, it will clearly be enough to prove the following lemma.

Lemma 5.7 *Let x be an atom in its diagonal, let $0 < \tau < 1$ and let $y = y^* \in X$ be so that $y = P_\tau y$. Then $y = 0$.*

The proof of the lemma is by contradiction and is long and technical. We shall first prove a special case of the lemma, and then show how the general case is reduced to the special case.

The special case. We assume that

$$x = \sum_{j=0}^{\infty} \tau^j p_j \tag{5.4}$$

and

$$y = \sum_{j=1}^{\infty} (a_j + a_j^*) \tag{5.5}$$

where $a_j = p_{j-1} a_j p_j$ and $a_1 \ne 0$.

Thus,

$$x = \begin{pmatrix} p_0 & 0 & 0 & 0 \\ 0 & \tau p_1 & 0 & 0 \\ 0 & 0 & \tau^2 p_2 & 0 \\ 0 & 0 & 0 & \ddots \end{pmatrix} \tag{5.6}$$

and

$$y = \begin{pmatrix} 0 & a_1 & 0 & 0 & \dots & 0 \\ a_1^* & 0 & a_2 & 0 & \dots & 0 \\ 0 & a_2^* & 0 & a_3 & \dots & 0 \\ 0 & 0 & a_3^* & 0 & \dots & 0 \\ \vdots & \vdots & \vdots & \vdots & \ddots & \\ 0 & 0 & 0 & 0 & & 0 \end{pmatrix}$$

Let us denote $f_r(\lambda) = sgn(\lambda)|\lambda|^r$, and apply f_r to operators via the triple functional calculus.

Let us assume first that $3 < p$. Then $t \to f_{p-1}(x + ty)$ is differentiable of order 2 (see Theorem 3.10), and

$$\frac{1}{2}\frac{d^2}{dt^2} f_{p-1}(x + ty)_{|t=0} = \sum_{0 \le i,j,k} f_{p-1}^{[2]}(\tau^j, \tau^i, \tau^k) p_j y p_i y p_k. \tag{5.7}$$

Note that $f_{p-1}(x+ty) \in R(P^*)$. Since $x^{p-1}/\|x\|_p^{p-1}$ is an atom in its diagonal (with respect to P^*), for every $t \ne 0$ there is a $\lambda(t) \in \mathbf{C}$ so that

$$\sum_{j=0}^{\infty} p_j[(f_{p-1}(x + ty) + f_{p-1}(x - ty)) - 2f_{p-1}(x))/2t^2]p_j$$

$$= \lambda(t) f_{p-1}(x) = \lambda(t) \sum_{j=0}^{\infty} \tau^{(p-1)j} p_j. \tag{5.8}$$

By (5.7), $\lambda(0) := \lim_{t \to 0} \lambda(t)$ exists and

$$\lambda(0) \sum_{j=0}^{\infty} \tau^{(p-1)j} p_j = \sum_{i,j=0}^{\infty} f_{p-1}^{[2]}(\tau^j, \tau^i, \tau^j) p_j y p_i y p_j. \tag{5.9}$$

Using (5.4) and (5.9) we get

$$\lambda(0) p_0 = f_{p-1}^{[2]}(1, \tau, 1) a_1 a_1^* \tag{5.10}$$

and

$$\lambda(0) \tau^{p-1} p_1 = f_{p-1}^{[2]}(\tau, 1, \tau) a_1^* a_1 + f_{p-1}^{[2]}(\tau, \tau^2, \tau) a_2 a_2^*. \tag{5.11}$$

As f_{p-1} is strictly convex, $f_{p-1}^{[2]}(\tau, \tau^2, \tau) > 0$. Thus by (5.11) and (5.10),

$$f_{p-1}^{[2]}(\tau, 1, \tau) \|a_1\|_\infty^2 \le |\lambda(0)| \tau^{p-1}$$

$$= \tau^{p-1} f_{p-1}^{[2]}(1, \tau, 1) \|a_1\|_\infty^2. \tag{5.12}$$

The assumption $a_1 \neq 0$ leads by (5.12) to the inequality $f_{p-1}^{[2]}(\tau, 1, \tau) \leq \tau^{p-1} f_{p-1}^{[2]}(1, \tau, 1)$. However, it is elementary to verify directly that $f_{p-1}^{[2]}(\tau, 1, \tau) > \tau^{p-1} f_{p-1}^{[2]}(1, \tau, 1)$. This contradiction completes the proof in case $3 < p$.

For a general $2 < p < \infty$, $\frac{d^2}{dt^2} f_{p-1}(x+ty)_{|t=0}$ need not exist. The proof is modified so as to overcome this difficulty. For $t \neq 0$ we have

$$\frac{f_{p+1}(x+ty) + f_{p+1}(x-ty) - 2f_{p+1}(x)}{2t^2} =$$

$$x\left(\frac{f_{p-1}(x+ty) + f_{p-1}(x-ty) - 2f_{p-1}(x)}{2t^2}\right)x + \tag{5.13}$$

$$y\left(\frac{f_{p-1}(x+ty) + f_{p-1}(x-ty)}{2}\right)y +$$

$$\left\{x, \frac{f_{p-1}(x+ty) - f_{p-1}(x-ty)}{t}, y\right\},$$

where $\{\cdot, \cdot, \cdot\}$ is the triple product. Notice that using Theorem 3.1 and (3.8)

$$\lim_{t \to 0} \frac{f_{p-1}(x+ty) - f_{p-1}(x-ty)}{t} = 2\frac{d}{dt} f_{p-1}(x+ty)_{|t=0}$$

$$= 2 \sum_{i,j=0}^{\infty} f_{p-1}^{[1]}(\tau^i, \tau^j) p_i y p_j. \tag{5.14}$$

Using the fact that $x^{p-1}/\|x\|_p^{p-1}$ is atom in its diagonal (for P^*), we get as in (5.8) above,

$$\sum_{j=0}^{\infty} p_j \left[x\left(\frac{f_{p-1}(x+ty) + f_{p-1}(x-ty) - 2_{p-1}(x)}{2t^2}\right)x\right] p_j$$

$$= x\left[\sum_{j=0}^{\infty} p_j \left(\frac{f_{p-1}(x+ty) + f_{p-1}(x-ty) - 2f_{p-1}(x)}{2t^2}\right) p_j\right] \tag{5.15}$$

$$= \lambda(t) x\, x^{p-1} x = \lambda(t) \sum_{j=0}^{\infty} \tau^{(p+1)j} p_j.$$

Using (5.13) and Theorem 3.10, we get that $\lambda(0) := \lim_{t \to 0} \lambda(t)$ exits. Moreover, by applying the diagonal projection $z \mapsto \sum_{j=0}^{\infty} p_j z p_j$ and using (5.14) we obtain

$$\lambda(0) \sum_{j=0}^{\infty} \tau^{(p+1)j} p_j = \sum_{i,j=0}^{\infty} f_{p+1}^{[2]}(\tau^j, \tau^i, \tau^j) p_j y p_i y p_j - \sum_{i,j=0}^{\infty} \tau^{(p-1)i} p_j y p_i y p_j$$

$$-2\sum_{j=0}^{\infty} p_j \left(\sum_{k,\ell=0}^{\infty} f_{p-1}^{[1]}(\tau^k, \tau^\ell)\{x, p_k y p_\ell, y\} \right) p_j \tag{5.16}$$

$$= \sum_{i,j=0}^{\infty} \left(f_{p+1}^{[2]}(\tau^j, \tau^i, \tau^j) - \tau^{i(p-1)} - 2\tau^j f_{p-1}^{[1]}(\tau^i, \tau^j) \right) p_j y p_i y p_j.$$

Comparing the $(0,0)$ and $(1,1)$ diagonal entries of both sides of (5.16) we get

$$\lambda(0)p_0 = \left(f_{p+1}^{[2]}(1, \tau, 1) - \tau^{p-1} - 2 f_{p-1}^{[1]}(\tau, 1) \right) a_1 a_1^* \tag{5.17}$$

and

$$\lambda(0)\tau^{p+1}p_1 = \left(f_{p+1}^{[2]}(\tau, 1, \tau) - 1 - 2\tau f_{p-1}^{[1]}(1, \tau) \right) a_1 a_1^*$$

$$+ \left(f_{p+1}^{[2]}(\tau, \tau^2, \tau) - \tau^{2(p-1)} - 2\tau f_{p-1}^{[1]}(\tau^2, \tau) \right) a_2 a_2^*. \tag{5.18}$$

However, a direct manipulation of the divided differences gives

$$f_{p+1}^{[2]}(1, \tau, 1) - \tau^{p-1} - 2 f_{p-1}^{[1]}(\tau, 1) = f_{p-1}^{[2]}(1, \tau, 1), \tag{5.19}$$

$$f_{p+1}^{[2]}(\tau, 1, \tau) - 1 - 2\tau f_{p-1}^{[1]}(1, \tau) = \tau^2 f_{p-1}^{[2]}(\tau, 1, \tau), \tag{5.20}$$

and

$$f_{p+1}^{[2]}(\tau, \tau^2, \tau) - \tau^{2(p-1)} - 2\tau f_{p-1}^{[1]}(\tau^2, \tau) = \tau^2 f_{p-1}^{[2]}(\tau, \tau^2, \tau). \tag{5.21}$$

Hence (5.17) and (5.19) imply (5.10) and (5.18), and (5.20) and (5.21) imply (5.11). The rest of the proof is as in the case $3 < p$ considered above, and the contradiction obtained fron (5.12) completes the proof of the special case in Lemma 5.7.

Reduction of the general case in Lemma 5.7 to the special case.

A *chain* is a sequence $j_1 < j_2 < \ldots$ (finite or infinite, in case $n = \infty$) so that $(j_\nu, j_{\nu+1}) \in A_\tau$ for all ν. Given a chain $J = \{j_\nu\}$, let $A_\tau(J) = \{(j_\nu, j_{\nu+1}); j_\nu \in J\}$. We call a chain *maximal* if it is not a proper subchain of any other chain.

It is easy to verify that

$$A_\tau = \cup_k A_\tau(J_k)$$

where $\{J_k\}$ is a family of maximal pairwise disjoint chains. If

$$y_k = \sum_{(i,j)\in A_\tau(J_k)} P_{i,j}y,$$

then $\{y_k\}$ are pairwise orthogonal. Also, $y = \sum y_k$. Fix k, and let $J_k = \{j_\nu\}_{0\le\nu<m}$ be the enumeration of J_k (where $2 \le m \le \infty$). Since $(j_\nu, j_{\nu+1}) \in A_\tau$ for every ν, we have $\alpha_{j_{\nu-1}}/\alpha_{j_\nu} = \tau$, and so $\alpha_{j_\nu} = \alpha_{j_0}\tau^\nu$. If $y \neq 0$ then for some k, $y_k \neq 0$. Clearly,

$$y_k = \sum(b_\nu + b_\nu^*), \quad b_\nu = p_{j_{\nu-1}}\, b_\nu p_{j_\nu}.$$

Let ν_0 be the first number so that $b_{\nu_0} \neq 0$. Let $a_\nu = b_{\nu+\nu_o-1}$, $1 \le \nu < m+1-\nu_o = \ell$.

Let

$$x_k = \sum_{\nu_0-1\le\nu<m} \alpha_{j_\nu}\alpha_{j_{\nu_0-1}}^{-1} p_{j_\nu} = \sum_{0\le\nu<\ell} \tau^\nu q_\nu$$

where $q_\nu = p_{j_{\nu+\nu_o-1}}$.

Notice that x_k is a multiple of a principal block of x, and that x_k and y_k look exactly as the x and y in the special case except that, in general, x_k and y_k do not belong to X. We now follow the proof of the special case, and pay attention to the picture in the k'th block. As in the special case, this leads to a contradiction. □

The proof of Lemma 5.7 is over, and this completes also the proofs of Lemma 5.4 and Theorem 5.2.

6. Basic relations between atoms

This section is devoted to the study of the possible relations between atoms of a contractive projection P on C_p $(1 < p < \infty, p \neq 2)$. Recall (Theorem 4.1) that if $0 \neq x \in R(P) = X$ has a polar decomposition $x = v(x)|x|$, then the Peirce projections $P_k(x) := P_k(\,v(x)\,)$ $(k = 0,1,2)$ commute with P and so $X_k(x) := P_k(x)X$ are subspaces of X which are the ranges of the contractive projections $P_k(x)P$, $k = 0,1,2$.

Definition 6.1 Let $x, y \in C_p$.

(i) x and y are *compatible* if their Peirce projections commute, i.e. the family $\{P_0(x), P_1(x), P_2(x), P_0(y), P_1(y), P_2(y)\}$ is commutative;

(ii) x and y are *colinear*, in notation $x \top y$, if there exist elements $a, b \in C_p$ with $\|a\|^p + \|b\|^p = 1$ and a matrix representation so that

$$x = \left(\begin{array}{cc|cc} 0 & a & 0 & 0 \\ b & 0 & 0 & 0 \\ \hline 0 & 0 & 0 & 0 \\ 0 & 0 & 0 & 0 \end{array}\right), \quad y = \left(\begin{array}{cc|cc} 0 & 0 & a & 0 \\ 0 & 0 & 0 & 0 \\ \hline b & 0 & 0 & 0 \\ 0 & 0 & 0 & 0 \end{array}\right); \tag{6.1}$$

(iii) y *governs* x, in notation $y \vdash x$, if there exist an element $a \in C_p$ with $\|a\| = 1$ and a matrix representation so that

$$x = \begin{pmatrix} a & 0 & 0 \\ 0 & 0 & 0 \\ 0 & 0 & 0 \end{pmatrix}, \quad y = 2^{-1/p} \begin{pmatrix} 0 & a & 0 \\ a & 0 & 0 \\ 0 & 0 & 0 \end{pmatrix}. \tag{6.2}$$

These relations are clearly the generalizations of the same relations between tripotents in JB^*-triples, see [DF]. Clearly, x and y are compatible if and only if so are $v(x)$ and $v(y)$. Also, if $x = P_k(y)x$ for some k, then x and y are compatible. So if $x \top y$ or $y \vdash x$ then x and y are compatible. Also, $x \top y$ (or $y \vdash x$) implies $v(x) \top v(y)$ (respectively, $v(y) \vdash v(x)$), but the converse is false in general.

We remark that the existence of a matrix representation (6.1) is equivalent to the existence of a tensor product representation in which

$$x = a \otimes e_{1,2} + b \otimes e_{2,1} \qquad \text{and} \qquad y = a \otimes e_{1,3} + b \otimes e_{3,1}. \tag{6.3}$$

Similarly, (6.2) is equivalent to

$$x = a \otimes e_{1,1}, \qquad y = 2^{-1/p}(a \otimes (e_{1,2} + e_{2,1})). \tag{6.4}$$

One observes that if $y \vdash x$ then $\|y\| = \|x\| = 1$, and if $x \top y$ then

$$\|\alpha x + \beta y\| = (|\alpha|^2 + |\beta|^2)^{1/2} \tag{6.5}$$

for all choices of scalars α, β.

Lemma 6.2 *(The "two-case lemma") Let x be an atom of $X = R(P)$ and let $y \in X_1(x), \|y\| = 1$. Then exactly one of the following cases must occur: either*

Case I : $x \in X_1(y)$ and x and y are colinear;

or

Case II: $x \in X_2(y)$, the element $\tilde{x}$ defined by

$$\tilde{x} = Q\,(\,v(y)\,)\,x \tag{6.6}$$

is an atom of X, and

$$X_0(y) \cap X_1(x) = \{0\} = X_0(y) \cap X_2(x). \tag{6.7}$$

Proof: Put $D(y) := D(v(y)) = \{v(y), v(y), \cdot\}$, so $D(y) = P_2(y) + \frac{1}{2}P_1(y)$. Since y is compatible with x, we get by Corollary 4.2

$$D(y)x = P_2(x)PD(y)x.$$

Since x is an atom of P there is a $\lambda \in \mathbf{C}$ so that $D(y)x = \lambda$x. Thus $\lambda \in \sigma\,(\,D(y)\,) \subseteq \{0, \frac{1}{2}, 1\}$. The case $\lambda = 0$ is impossible since it would mean that x and y are orthogonal, contradicting $y \in X_1(x)$. Thus either Case I : $\lambda = \frac{1}{2}$ and $x \in X_1(y)$, or, Case II: $\lambda = 1$ and $x \in X_2(y)$.

Assume first that we are in Case I. We claim that y is an atom of $X_1(x)$. If not, then by Theorem 5.2 y is decomposable, thus $y = y_1 + y_2$, where y_1, y_2 are non-zero orthogonal elements of $X_1(x)$. The argument in the beginning of the proof yields $D(y_j)x = \lambda_j x$ with $\lambda_j \in \{\frac{1}{2}, 1\}, j = 1, 2$. Now, $D(y_1+y_2) = D(y_1)+D(y_2)$ since $y_1 \perp y_2$, so

$$\frac{1}{2}x = D(y)x = D(y_1 + y_2)x = D(y_1)x + D(y_2)x = (\lambda_1 + \lambda_2)x$$

Thus $\lambda_1 + \lambda_2 = \frac{1}{2}$, a contradiction which proves that y is an atom of $X_1(x)$.

Since $x \in X_1(y)$ and $y \in X_1(x)$ there are decompositions $x = x_1 + x_2$ and $y = y_1 + y_2$, so that the left and right support projections satisfy: $r(x_1) \perp r(y_1)$, $l(x_1) =$

$l(y_1)$, $r(x_2) = r(y_2)$ and $l(x_2) \perp l(y_2)$. This leads to a matrix representation in which

$$x = \left(\begin{array}{cc|cc} 0 & x_1 & 0 & 0 \\ x_2 & 0 & 0 & 0 \\ \hline 0 & 0 & 0 & 0 \\ 0 & 0 & 0 & 0 \end{array}\right), \; y = \left(\begin{array}{cc|cc} 0 & 0 & y_1 & 0 \\ 0 & 0 & 0 & 0 \\ \hline y_2 & 0 & 0 & 0 \\ 0 & 0 & 0 & 0 \end{array}\right). \tag{6.8}$$

We assume without loss of generality that $p > 2$. Applying Proposition 3.8, Corollary 3.9 and (6.8) we get that

$$z := N(x)y = \frac{d}{dt}N_p(x+ty)_{|t=0} \in R(P^*)$$

has a matrix

$$z = \left(\begin{array}{cc|cc} 0 & 0 & |x_1^*|^{p-2}y_1 & 0 \\ 0 & 0 & 0 & 0 \\ \hline y_2|x_2|^{p-2} & 0 & 0 & 0 \\ 0 & 0 & 0 & 0 \end{array}\right) = \left(\begin{array}{cc|cc} 0 & 0 & z_1 & 0 \\ 0 & 0 & 0 & 0 \\ \hline z_2 & 0 & 0 & 0 \\ 0 & 0 & 0 & 0 \end{array}\right).$$

Notice that $l(z_j) = l(y_j)$ and $r(z_j) = r(y_j), j = 1,2$. Since $N_p(y)$ is an atom of $P_1(N_p(x))P^*$, there is a scalar c so that $z = cN_p(y)$.

Thus

$$|x_1^*|^{p-2}y_1 = c|y_1^*|^{p-2}y_1,$$

$$y_2|x_2|^{p-2} = cy_2|y_2|^{p-2}$$

and so

$$|x_1^*|^{p-2}|y_1^*|^2|x_1^*|^{p-2} = |c|^2|y_1^*|^{2(p-1)},$$

$$|x_2|^{p-2}|y_2|^2|x_2|^{p-2} = |c|^2|y_2|^{2(p-1)}.$$

These relations, together with $\|x\| = \|y\| = 1$, imply that

$$|x_1^*| = |y_1^*|\,,\; |x_2| = |y_2| \text{ and } c = 1. \tag{6.9}$$

The colinearity of x and y is now an easy consequence of (6.8), (6.9) and the following standard result whose proof is omitted.

Proposition 6.3 *Let $x, y \in B(H)$ be so that $\ell(x) \perp \ell(y)$ and $|x| = |y|$. Then there is a matrix representation and an element $a \in B(H)$ so that*

$$x = \begin{pmatrix} a & 0 \\ 0 & 0 \end{pmatrix}, \; y = \begin{pmatrix} 0 & 0 \\ a & 0 \end{pmatrix}.$$

We continue the proof of Lemma 6.2 in Case II, i.e. $x \in X_2(y)$. Being an atom of X, x is clearly an atom of $X_2(y)$. The operator $Q = Q(v(y))$ is a conjugate-linear automorphism of $X_2(y)$ satisfying $Q^2 = P_2(y) = I|_{X_2(y)}$, see section 2. Thus $\tilde{x} := Qx$ is also an atom of $X_2(y)$. This implies, via the following easy result, that $\tilde{x}$ is an atom of X.

Proposition 6.4 *Let P be a contractive projection in C_p $(1 < p < \infty, p \neq 2)$ and let e_1 and e_2 be projections in $B(H)$ so that the projection $Ez = e_1 z e_2$ commutes with P. Then every atom of EP is an atom of P. In particular, if $y \in R(P)$ and z is an atom of $P_2(y)P$ or $P_0(y)P$, then z is an atom of P.*

We remark that Proposition 6.4 need not be true for $P_1(y)P$ in place of $P_2(y)P$ or $P_0(y)P$. More generally, it does not hold if $Ez = e_1 x e_2 + e_3 x e_4$ where e_j are non-zero projections with $e_1 \perp e_3$ and $e_2 \perp e_4$.

To complete the proof of Lemma 6.2 it remains to verify (6.7). If $0 \neq \tilde{y} \in X_0(y) \cap X_1(x)$, then there exists $\lambda \in \{\frac{1}{2}, 1\}$ such that $D(y+\tilde{y})x = D(y)x + D(\tilde{y})x = (1+\lambda)x$, contradicting the fact that $\sigma(D(y+\tilde{y})) \subseteq \{0, \frac{1}{2}, 1\}$. If $0 \neq \tilde{y} \in X_0(y) \cap X_2(x)$, then $\tilde{y} = \lambda x$ with $\lambda \neq 0$ (since x is an atom) and so $\tilde{y} \in X_2(y)$(since $x \in X_2(y)$), a contradiction. This completes the proof of Lemma 6.2. □

In case II of Lemma 6.2, y is clearly not an atom of P since $\dim X_2(y) \geq 3$. In fact, y need not even be an atom of $X_1(x)$ and need not govern x. For example, let $x = e_{1,1}$, and $y = \alpha e_{1,2} + \beta e_{2,1}$ $(0 < \alpha, \beta < 1, \alpha \neq \beta, \alpha^p + \beta^p = 1)$.

However in Case I, y must be an atom of P. This follows from the following result, generalizing the "*Colinear exchange theorem*" of K. McCrimmon [Mc].

Lemma 6.5 *Let x be an atom of $X = R(P)$, let $y \in X_1(x)$ be so that $\|y\| = 1$ and $x \in X_1(y)$, and let $z = \alpha x + \beta y$ with $\alpha, \beta \in \mathbf{R}$ and $\alpha^2 + \beta^2 = 1$. Then there is a linear isometry $\Phi = \Phi_{x,z}$ of C_p onto itself so that*

(i) $\Phi^2 = I$, i.e. Φ is a symmetry;

(ii) $\Phi P = P\Phi$ and $\Phi(X) = X$;

(iii) $\Phi x = z$ and $\Phi z = x$;

(iv) $\Phi w = w$ for every $w \in X_0(x) \cap X_0(z)$.

Lemma 6.5 has the following corollary.

Corollary 6.6 *Let x and y be as in Lemma 6.5. Then y is an atom of P. Moreover, every normalized element in the complex linear span of x and y is an atom of P.*

Proof: Lemma 6.5 with $\alpha = 0, \beta = 1$ yields that y is an atom of P. Thus $e^{i\theta}x$ and $e^{it}y$ are atoms of P. Every normalized complex linear combination of x and y is of the form $z = \alpha e^{i\theta}x + \beta e^{it}y$, with α, β real and $\alpha^2 + \beta^2 = 1$, and $\theta, t \in \mathbf{R}$. Thus z is an atom of P, by Lemma 6.5 with $e^{i\theta}x$ and $e^{it}y$ in place of x and y respectively. □

Proof of Lemma 6.5 By Lemma 6.2 there is a tensor product representation and $a, b \in C_p$ with $\|a\|^p + \|b\|^p = 1$ and $a \perp b$, so that

$$x = a \otimes e_{0,1} + b \otimes e_{1,0} \qquad \text{and} \qquad y = a \otimes e_{0,2} + b \otimes e_{2,0}. \tag{6.9}$$

If $z = \pm x$, we define $\Phi = \pm I$ and obtain (i)-(iv). So assume that $z = \alpha x + \beta y$ is not proportional to x, i.e. $\beta \neq 0$. Define

$$u = \frac{x+z}{\|x+z\|}, v = v(u) \tag{6.10}$$

and set

$$\Phi = e^{2\pi i D(v)} = P_2(v) - P_1(v) + P_0(v). \tag{6.11}$$

Clearly, Φ is a surjective isometry of C_p and (i) and (ii) hold. To prove (iii) let $\tilde{u} = (x-z)/\|x-z\|$. Then by (6.9) u and $\tilde{u}$ are colinear. Hence $\tilde{u} \in X_1(v)$ and so $\Phi(u) = u$ and $\Phi(\tilde{u})$=-ũ. (iii) follows from this and the formulas:

$$x = \frac{\|x+z\|}{2}u + \frac{\|x-z\|}{2}\tilde{u}$$

$$z = \frac{\|x+z\|}{2}u - \frac{\|x-z\|}{2}\tilde{u}.$$

Let $w \in X_0(x) \cap X_0(z)$; then $P_0(w)x = x$ and $P_0(w)z$=z. Thus $P_0(w)u = u$ as well. Hence $u \perp w, P_0(u)w = w$ and

$$\Phi w = (P_2(u) - P_1(u) + P_0(u))w = P_0(u)w = w.$$

This completes the proof. □

We continue the analysis in Case II of Lemma 6.2.

A family of normalized elements in C_p is said to be *colinear* if every two elements from the family are colinear.

Corollary 6.7. *Let $\{x_j\}_{j\in J}$ be a colinear family of atoms of P.*

(i) *$\{x_j\}_{j\in J}$ is isometrically equivalent to an orthonormal sequence in a Hilbert space, i.e., $\|\sum_{j\in J}\alpha_j x_j\| = (\sum_{j\in J}|\alpha_j|^2)^{1/2}$ for all choices of scalars α_j;*

(ii) *If $J_0 \subset J$ is a finite subset and $z = \sum_{j\in J_0}\alpha_j x_j$ is normalized, then z is an atom of P and is colinear to any element x_j with $j \in J \setminus J_0$.*

The proof of part(ii) as well as the formula $\|\sum_{j\in J_0}\alpha_j x_j\| = (\sum_{j\in J_0}|\alpha_j|^2)^{1/2}$ follow from Lemma 6.5 and Corollary 6.6 by induction on the cardinality of J_0. Part (i) follows from the fomula $\|\sum_{j\in J_0}\alpha_j x_j\| = (\sum_{j\in J_0}|\alpha_j|^2)^{1/2}$ for finite subsets J_0 of J.

Remark 6.8. Colinear families in C_p ($p = 1, \infty$) were considered in [AF2, Def. 5.13, p. 92]. In [AF2] the colinearity condition is called the G-condition, and a colinear family is a family satisfying the strong G-condition. We remark also that, with the results of Section 7, it follows that if $\{x_j\}_{j\in J}$ is a colinear family of atoms of P, then every normalized element in $\overline{span}\{x_j\}_{j\in J}$ is an atom and two normalized elements $a = \sum_{i\in J}\alpha_j x_j, , b = \sum_{j\in J}\beta_j x_j$ are colinear if and only if $\sum_{j\in J}\alpha_j\bar{\beta}_j = 0$.

Lemma 6.9 *Let x and y be as in Case II of Lemma 6.2, i.e. x and y are normalized elements of $X = R(P)$, x an atom of $X, y \in X_1(x)$ and $x \in X_2(y)$. Assume, further, that y is an atom of $X_1(x)$. Then y governs x. Thus, there exists a tensor product representation and a normalized element $a \in C_p$ so that (6.4) holds, and the element $\tilde{x}$ defined by (6.6) is given by*

$$\tilde{x} = a \otimes e_{2,2}. \tag{6.12}$$

Moreover, any normalized element $z \in X_1(x) \cap X_1(y)$ is an atom of $X_1(x)$ which governs x and is colinear with y. Thus the above tensor product representation can be extended so that

$$z = 2^{-1/p}(a \otimes (e_{1,3} + e_{3,1})) \tag{6.13}$$

Proof: By Lemma 6.2 there exists a matrix representation in which

$$x = \begin{pmatrix} a & 0 & 0 \\ 0 & 0 & 0 \\ 0 & 0 & 0 \end{pmatrix}, \; y = \begin{pmatrix} 0 & y_1 & 0 \\ y_2 & 0 & 0 \\ 0 & 0 & 0 \end{pmatrix}, \tilde{x} = \begin{pmatrix} 0 & 0 & 0 \\ 0 & \tilde{a} & 0 \\ 0 & 0 & 0 \end{pmatrix},$$

where $\ell(a) = \ell(y_1)$, $r(a) = r(y_2)$, $r(y_1) = r(\tilde{a})$ and $\ell(y_2) = \ell(\tilde{a})$. As in the proof of Lemma 6.2, Case I, we see that

$$w := N_p'(x)y = \frac{d}{dt}N_p(x + ty)_{|t=0}$$

belongs to $Y := R(P^*)$, and has a matrix

$$w = \begin{pmatrix} 0 & w_1 & 0 \\ w_2 & 0 & 0 \\ 0 & 0 & 0 \end{pmatrix}$$

with

$$w_1 = |a^*|^{p-2} y_1, \qquad w_2 = y_2 |a|^{p-2}.$$

Next, $u := N_p(y)$ is an atom of $Y_1(\, N_p(x)\,)$ and

$$u = \begin{pmatrix} 0 & u_1 & 0 \\ u_2 & 0 & 0 \\ 0 & 0 & 0 \end{pmatrix}$$

with

$$u_1 = |y_1^*|^{p-2} y_1, \qquad u_2 = y_2 |y_2|^{p-2}.$$

Thus $w = cu$ for some scalar c. The proof of Case I in Lemma 6.2 shows that

$$|y_1^*| = 2^{-1/p} |a^*|, \qquad |y_2| = 2^{-1/p} |a|.$$

Using Proposition 6.3, we find a tensor product representation in which (6.4) holds. Using the definition of $\tilde{x}$, i.e. (6.6), we see that (6.12) holds.

Let z be a normalized element of $X_1(x) \cap X_1(y)$. Thus, in the above matrix representation

$$z = \begin{pmatrix} 0 & 0 & z_1 \\ 0 & 0 & 0 \\ z_2 & 0 & 0 \end{pmatrix},$$

with $l(z_1) \leq l(y_1)$ and $r(z_2) \leq r(y_2)$. To prove that z and y are colinear, we use Lemma 6.2 with y and z in place of x and y respectively, and $P_1(x)$ in place of P. If z and y are not colinear, then we are in Case II of Lemma 6.2. So $\tilde{y} := Q(v(z))y$ is an atom of $P_1(x)$. But by formula (2.10) (i.e. "Peirce arithmetic") we see that $\tilde{y} \in X_0(y) \cap X_1(x)$, contradicting (6.7). Thus z and y are colinear.

The fact that z is an atom of $X_1(x)$ follows from the colinearity of z and y, the fact that y is an atom of $X_1(x)$ and Lemma 6.2, Case I, with $P_1(x)P$ in place of P.

The colinearity of y and z enables us to extend the tensor product representation so that (6.13) holds. This together with (6.4) yields that z governs x. □

The last result of this section completes the analysis of Case II in Lemma 6.2; it deals with the case where y is not an atom of $X_1(x)$.

Lemma 6.10 *Let x and y be as in Case II of Lemma 6.2, i.e. x and y are normalized elements of $X = R(P), x$ is an atom of X, $y \in X_1(x)$ and $x \in X_2(y)$. Suppose that y is not an atom of $X_1(x)$. Then there exist normalized orthogonal elements $z, \tilde{z}$ of $X_1(x)$ and non-zero scalars $c, \tilde{c}$ so that*

$$y = cz + \tilde{c}\tilde{z}, \ x \in X_1(z) \cap X_1(\tilde{z}). \tag{6.14}$$

Let $\tilde{x} = Q(v(y))x$. Then there exists a tensor product representation and orthogonal elements a, b of C_p with $\|a\|^p + \|b\|^p = 1$, *so that*

$$x = a \otimes e_{1,1} + b \otimes e_{1,1}; \qquad \tilde{x} = a \otimes e_{2,2} + b \otimes e_{2,2};$$

$$z = a \otimes e_{1,2} + b \otimes e_{2,1}; \qquad \tilde{z} = a \otimes e_{2,1} + b \otimes e_{1,2}. \tag{6.15}$$

Moreover, $z, \tilde{z}$ and $\tilde{x}$ are atoms of X.

The situation described in (6.15) can be written in matrix form as follows:

$$x = \left(\begin{array}{cc|cc} a & 0 & 0 & 0 \\ 0 & 0 & 0 & 0 \\ \hline 0 & 0 & b & 0 \\ 0 & 0 & 0 & 0 \end{array}\right); \ \tilde{x} = \left(\begin{array}{cc|cc} 0 & 0 & 0 & 0 \\ 0 & a & 0 & 0 \\ \hline 0 & 0 & 0 & 0 \\ 0 & 0 & 0 & b \end{array}\right) \tag{6.16a}$$

$$z = \left(\begin{array}{cc|cc} 0 & a & 0 & 0 \\ 0 & 0 & 0 & 0 \\ \hline 0 & 0 & 0 & 0 \\ 0 & 0 & b & 0 \end{array}\right); \ \tilde{z} = \left(\begin{array}{cc|cc} 0 & 0 & 0 & 0 \\ a & 0 & 0 & 0 \\ \hline 0 & 0 & 0 & b \\ 0 & 0 & 0 & 0 \end{array}\right) \tag{6.16b}$$

Proof: Since y is not an atom of $X_1(x)$, we get by Theorem 5.2 that y is decomposable. Thus there exist orthogonal normalized elements $z, \tilde{z}$ of $X_1(x)$ and non-zero scalars $c, \tilde{c}$ so that $y = cz + \tilde{c}\tilde{z}$. As in Lemma 6.2, $D(v(z))x = \lambda x$ and $D(v(\tilde{z})) = \tilde{\lambda} x$ where $\lambda, \tilde{\lambda} \in \{\frac{1}{2}, 1\}$. But $v(y) = v(z) + v(\tilde{z})$, and so by the orthogonality of $z, \tilde{z}$,

$$x = D(\, v(z)\,)\,x = D(\, v(z)\,)\,x + D(\, v(\tilde{z})\,)\,x = (\lambda + \tilde{\lambda})x.$$

This means that $\lambda = \tilde{\lambda} = \frac{1}{2}$ and so $x \in X_1(z) \cap X_1(\tilde{z})$ and (6.14) holds. By Case I of Lemma 6.2, the pairs $\{x, z\}$ and $\{x, \tilde{z})$ are colinear. This leads to the existence of a tensor product representation and orthogonal elements $a, b \in C_p$ with $\|a\|^p + \|b\|^p = 1$, so that (6.15) holds for x, z and $\tilde{}z$. Since $v(y) = v(z) + v(\tilde{z})$, we see that $\tilde{x} = Q(\, v(y)\,)\,x$ is given by (6.15) as well. Notice that $\tilde{x}$ is colinear with z and $\tilde{}z$. The fact that $x, \tilde{z}$ and $\tilde{x}$ are atoms of X follows from Corollary 6.6 and the fact that the pairs $\{x, z\}, \{x, \tilde{z}\}$ and $\{z, \tilde{x}\}$ are colinear. □

The *rank* of a subspace X of C_p is the maximal cardinality of a family of pairwise orthogonal, non-zero elements of X.

Corollary 6.11 *Let x be an atom of $X = R(P)$. Then the rank of $X_1(x)$ is less than or equal to 2.*

This follows from Lemma 6.10.

Remark 6.12 Let $x, z, \tilde{x}$ and $\tilde{z}$ be given by (6.15). Then, according to the terminology of [AF2, p.104], $(x, z, \tilde{x}, -\tilde{z})$ satisfy the *matrix condition* (or *condition M* for short).

7. The structure of N-convex subspaces of C_p.

In order to describe the structure of an N-convex subspace of C_p ($1 < p < \infty, p \neq 2$) (or equivalently, a subspace which is the range of a contractive projection) it is enough by Proposition 2.2 to describe only the indecomposable ones. In this section we prove Theorem 2.4, in which the description of the indecomposable N-convex subspaces of C_p is given. Any such subspace X is isometric to a p-ideal U_p (defined by (2.19)) of a classical Cartan factor U and the isometry T of U_p onto X is in fact a triple isometry and is given by (2.26).

Since our description depends heavily on the type of the factor U, the proofs will differ from type to type. We shall prove Theorem 2.4 under various additional assumptions (leading to a definite type of the Cartan factor U). At the end of the section we shall prove that any indecomposable N-convex subspace of C_p satisfies one of these additional assumptions. This will complete the proof of Theorem 2.4.

Case 1: $U = U(III_n)$, the symmetric n by n matrices.

Theorem 7.1 *Let X be an indecomposable N-convex subspace of C_p $(1 < p < \infty, p \neq 2)$. Assume that there exist an atom x of X, an atom y of $X_1(x)$ so that $x \in X_2(y)$. Then there is an index set J, a family $\{x_{i,j}\}_{i,j \in J}$ of elements of X, an element $a \in C_p$ with $\|a\| = 1$, and a tensor product representation such that*

$$x_{i,j} = x_{j,i} = a \otimes 2^{-1/p}(e_{i,j} + e_{j,i}),\ i \neq j;\ i,j \in J; \tag{7.1}$$

$$x_{i,i} = a \otimes e_{i,i}, \quad i \in J; \tag{7.2}$$

and

$$X = \overline{\text{span}}\{x_{i,j};\ i,j \in J\}. \tag{7.3}$$

Here $e_{i,j} = (\cdot, e_j)e_i$, *where* $\{e_j\}_{j \in J}$ *is an orthonormal sequence.*

Moreover, let

$$U = \overline{\text{span}}^{w^*}\{e_{i,j} + e_{j,i};\ i,j \in J\}. \tag{7.4}$$

Then $U = U(III_n)$, *and*

$$X = \{a \otimes u; u \in U_p\} \tag{7.5}$$

where $U_p = U \cap C_p$ *is the p-ideal of U.*

Proof: Choose a maximal colinear family of normalized elements $(x_{1,j}\}_{3 \leq j \leq n}$ (where $n \leq \infty$) in $X_1(x) \cap X_1(y)$. Denote $x_{1,1} = x,\ x_{1,2} = y$ and $J = \{j; 1 \leq j \leq n\}$. By

Lemma 6.9 there is an element $a \in C_p$, $\|a\| = 1$ and a tensor product representation such that (7.1) holds for $i = 1, j \in J, j \neq 1$, and (7.2) holds for $x_{1,1}$. Moreover, the $x_{1,j}$ with $j \neq 1$ are atoms of $X_1(x_{1,1})$.

For any $i \in J,\ i \neq 1$, define

$$x_{i,i} = Q\,(\,v(x_{1,i})\,)\,x_{1,1}$$

By Lemma 6.2, $x_{i,i}$ is an atom of X. To see that $x_{i,i}$ satisfies (7.2), note that $v(x_{1,i}) = v(a) \otimes (e_{1,i} + e_{i,1})$. Thus

$$\begin{aligned} x_{i,i} &= Q(v(a) \otimes (e_{1,i} + e_{i,1}))(a \otimes e_{1,1}) \\ &= Q(v(a))a \otimes Q(e_{1,i} + e_{i,1})e_{1,1} = a \otimes e_{i,i}. \end{aligned}$$

For any $i, j \in J\backslash\{1\},\ i \neq j$, define $x_{i,j}$ by

$$x_{i,j} = Q\,(\,v(x_{1,j} + x_{i,i})\,)\,x_{1,i}. \tag{7.6}$$

From Theorem 4.1(iii) it follows that $x_{i,j} \in X$ and, as above, one easily verifies that (7.1) holds for all the $x_{i,j}$. To see that $x_{i,j}$ is an atom of $X_1(x_{j,j})$, note first that the mapping $Q\,(\,v(x_{1,j} + x_{i,i})\,)$ is a conjugate-linear triple automorphism of $X_2(x_{1,j} + x_{i,i})$, mapping $x_{1,1}$ to $x_{j,j}$. Moreover, $Q(v(x_{1,j} + x_{i,i}))$ commutes with the Peirce projections of the $\{x_{k,l}\}$, and therefore it maps the subspace $X_1(x_{1,1}) \cap X_2(x_{1,i})$ onto $X_1(x_{j,j}) \cap X_2(x_{i,j})$. Since $x_{1,i}$ is an atom of $X_1(x_{1,1})$, the subspace $X_1(x_{1,1}) \cap X_2(x_{1,i})$ is one-dimensional. Thus $X_1(x_{j,j}) \cap X_2(x_{i,j})$ is one dimensional as well and $x_{i,j}$ is an atom of $X_1(x_{j,j})$.

To prove (7.3), consider the joint Peirce decomposition of X relative to the orthogonal family of tripotents $v_j = v(x_{j,j}), j \in J$, analogous to (2.15), namely

$$X = \sum_{0 \leq i \leq j \leq n} \oplus X_{i,j}$$

where

$$X_{j,j} = X_2(v_j), \qquad 1 \leq j \leq n;$$

$$X_{i,j} = X_1(v_i) \cap X_1(v_j), \qquad 1 \le i < j \le n;$$

$$X_{0,0} = \bigcap_{1 \le i \le n} X_0(v_i);$$

and

$$X_{0,j} = X_1(v_j) \cap \bigcap_{\substack{1 \le i \le n \\ i \ne j}} X_0(v_i).$$

Notice that $X_{i,j} = \mathbf{C}x_{i,j}$ for $1 \le i \le j \le n$. For $i = j$ this follows from the fact that $x_{i,i}$ is an atom of X. For $i < j$, this follows from the fact that $x_{i,j}$ is an atom of both $X_1(v_i)$ and $X_1(v_j)$.

We claim that $X_{0,j} = \{0\}$ for every $1 \le j \le n$. Indeed, suppose that $X_{0,j} \ne \{0\}$ for some $1 \le j \le n$. $X_{0,j}$ is the range of the contractive projection $P_{0,j}P$ (where $P_{0,j}$ is defined in (2.14)) and thus it has an atom which we denote by z, see Theorem 5.3. If $j = 1$ then by Lemma 6.9, z is colinear with $x_{1,k}$ for each $2 \le k \le n$, contradicting the maximality of the family $\{x_{1,k}\}$. Thus $X_{0,1} = \{0\}$. If $2 \le j \le n$, let $w = Q(v(z))x_{j,j}$. Using Lemma 6.9 again we see that w is an atom of X, which is orthogonal to $x_{i,i}$ for every $1 \le i \le n$. But then the element $Q(\, v(x_{1,j} + w)\,)\, z$ is an atom of $X_{0,1}$, a contradiction to the fact $X_{0,1} = \{0\}$ established above. Thus $X_{0,j} = \{0\}$ for all $1 \le j \le n$.

Since X is indecomposable, it follows that $X_{0,0} = \{0\}$ as well. Thus

$$X = \sum_{1 \le i \le j \le n} X_{i,j} = \sum_{1 \le i \le j \le n} \mathbf{C}x_{i,j},$$

which establishes (7.3).

The fact that the space U defined by (7.4) is a Cartan factor of type III_n follows from the definition of Cartan factors. It is obvious that the map $T : U_p \to X$ defined by $T(u) = a \otimes u$ is an isometry of U_p onto $\overline{\mathrm{span}}(x_{i,j}\}_{1 \le i \le j \le n} = X$.
Thus (7.5) holds. □

Case 2: $U = U(I_{1,n})$, the Hilbert spaces.

In this case $U_p(I_{1,n})$ is a Hilbert space. Its triple isometries into C_p differ from the triple isometries of $U_p(I_{n,m})$ with $\min\{m, n\} \ge 2$ into C_p (which will be considered later). We recall some results of [$AF2$]. These results are proved there in the context of C_1, but the same proofs establish them in C_p for $1 < p < \infty$, $p \neq 2$.

A finite or infinite sequence $\{x_j\}$ of normalized elements of C_p is said to be *strongly colinear* (see [$AF2$, Definition 5.13], where we used the terminology "strong G condition") if for each k, x_k is colinear to every normalized element in span $\{x_j\}_{j=1}^{k-1}$. Notice that from the Definition 6.1(ii) it follows that if $\{x_j\}$ are strongly colinear then

$$\|\sum_j t_j x_j\| = (\sum_j |t_j|^2)^{1/2}$$

for every choice of scalars $\{t_j\}$.

There is a complete description of the strongly colinear sequences in C_p in terms of certain elements $x_{n,k,m}$ of C_p defined in [$AF2$, p.16]. In particular, it follows that the notion of strong colinearity is independent of the ordering of the sequence $\{x_j\}$.

Lemma 7.2 *[AF2 ; Corollary 5.18] Let $\{x_k\}_{k=1}^n, 1 \leq n \leq \infty$, be a strongly colinear sequence in $C_p, 1 \leq p < \infty, p \neq 2$.*

(i) If $n < \infty$ then there exist elements $a_1, ..., a_n \in C_p$ satisfying

$$\sum_{m=1}^{n} \binom{n-1}{m-1} \|a_m\|^p = 1, \tag{7.7}$$

and there exists a tensor product representation so that for every $1 \leq k \leq n$:

$$x_k = h_{n,k}^{(p)}(a_1, ..., a_n) := \sum_{m=1}^{n} a_m \otimes x_{n,k,m}. \tag{7.8}$$

(ii) If $n = \infty$, then there exist $a_1, a_\infty \in C_p$ with $\|a_1\|^p + \|a_\infty\|^p = 1$, and there exists a tensor product representation in which for $1 \leq k < \infty$

$$x_k = h_{n,k}^{(p)}(a_1, a_\infty) = a_1 \otimes x_{\infty,k,1} + a_\infty \otimes x_{\infty,k,\infty}. \tag{7.9}$$

While the description of the elements $x_{n,k,m}$ is complicated (see [$AF2$], (2.15)), all we need here is the next lemma. Recall that two triple isometries T_1, T_2 from ℓ_2^n into $B(H)$ are said to be *equivalent* if there exist partial isometries v, w so that $T_1(x) = vT_2(x)w$ for every $x \in \ell_2^n$.

Lemma 7.3 *(i) For any $1 \leq m \leq n < \infty$ the map $T_{m,n} : \ell_2^n \to B(H)$ defined by*

$$T_{m,n}(e_j) = x_{n,j,m}, 1 \leq j \leq n \tag{7.10}$$

is a triple-monomorphism;

(ii) For $1 \leq k, m \leq n$ with $k \neq m$, $T_{k,n}$ is not equivalent to $T_{m,n}$;

(iii) The maps $T_\nu : H \to B(\ell_2)$, $\nu = 1, \infty$, defined by

$$T_\nu(e_k) = x_{\infty,k,\nu}, \qquad k = 1, 2 \ldots \tag{7.11}$$

are triple-monomorphisms. Moreover, T_1 is not equivalent to T_∞.

The case of $U = U(I_{1,n})$ in Theorem 2.4 is the following result.

Theorem 7.4 *Let X be an indecomposable N-convex subspace of C_p $(1 < p < \infty, p \neq 2)$. Assume that there exist colinear atoms x, y in X so that rank$(X_1(x)) = 1$. Then there exists a strongly colinear family $\{x_j\}_{j \in J}$ so that $X = \overline{\text{span}}\{x_j\}_{j \in J}$.*

If $n := |J| < \infty$, then there exist elements $\{a_m\}_{m=1}^n$ in C_p satisfying (7.7) and there is a tensor product representation so that (7.8) holds. Thus, the map $T : \ell^n \to X$ defined by:

$$Tu = \sum_{m=1}^{n} a_m \otimes T_{m,n}(u), \qquad u \in \ell_2^n; \tag{7.12}$$

is a surjective triple-isometry.

If $|J| = \infty$, then there exist $a_1, a_\infty \in C_p$ with $\|a_1\|^p + \|a_\infty\|^p = 1$, and a tensor product representation so that (7.9) holds. In this case the map $T : \ell_2 \to X$ defined via

$$Tu = a_1 \otimes T_1(u) + a_\infty \otimes T_\infty(u) \tag{7.13}$$

is a surjective triple-isometry.

In particular, X is isometric to a Hilbert space.

Remark: In formula (7.12) the spaces $a_m \otimes T_{m,n}(\ell_2^n), 1 \leq m \leq n$, are mutually orthogonal. Similarly, in (7.13) the space, $a_1 \otimes T_1(\ell_2)$ and $a_\infty \otimes T_\infty(\ell_2)$ are orthogonal. Thus the space X in Theorem 7.4 is triple-isometric to a Hilbert space and the formulas (7.12) or (7.13) give the decomposition of the triple-isometry T into its *irreducible components.*

Proof: By Zorn's lemma there is a maximal colinear family $\{x_j\}_{j\in J}$ of atoms of X, where J is either $\{1,2,\ldots,n\}$ or $\{1,2,\ldots,n,\ldots\}$, so that $x_1 = x$ and $x_2 = y$.

If $\bigcap_{j\in J} X_1(x_j) \neq \{0\}$, choose a normalized element z in this space. By Lemma 6.2 either z is colinear to x_1, or $x_1 \in X_2(z)$. If z is colinear to x_1 them by Corollary 6.6, z is an atom of X and is colinear to x_j for each $j \in J$. This contradicts the maximality of the family $\{x_j\}_{j\in J}$. If $x_1 \in X_2(z)$, z is not an atom and therefore by Lemma 6.2, $y \in X_2(z)$. But then, the non-zero element $Q(v(z))y$ belongs to $X_0(y) \cap X_1(x)$, contradicting the assumption that $\mathrm{rank}(X_1(x)) = 1$. Thus

$$\bigcap_{j\in J} X_1(x_j) = \{0\}. \tag{7.14}$$

By Lemma 6.5 any x_j can be interchanged with x by an automorphism. This automorphism interchanges the subspaces $X_1(x_j)$ and $X_1(x)$. Therefore $\mathrm{rank}(X_1(x_j)) = 1$ for any $j \in J$. Moreover, for any $i, j \in J, i \neq j$:

$$X_0(x_i) \cap X_1(x_j) = \{0\}. \tag{7.15}$$

Using the joint Peirce decomposition of X with respect to the family $\{x_j\}$, formulas (7.14) and (7.15) and the the fact that x_j is an atom, we get

$$X = \sum_{j\in J} X_2(x_j) + \bigcap_{j\in J} X_0(x_j) = \sum_{j\in J} \mathbf{C}x_j + \bigcap_{j\in J} X_0(x_j).$$

Since X is indecomposable, $\bigcap_{j\in J} X_0(x_j) = 0$. Thus $X = \sum_{j\in J} \mathbf{C}x_j$.

We claim now that the sequence $\{x_j\}_{j\in J}$ is strongly colinear. In fact, if $I \subset J$ is a finite subset then any normalized element z in span $\{x_j\}_{j\in I}$ is an atom of X which is colinear with x_k for any $k \in J\backslash$I. If $|I| = 2$, then z is an atom of X by Corollary 6.6. Since $x_j \in X_1(x_k)$ for $j \in I$, we get $z \in X_1(x_k)$. Since z is an atom of X, $x_k \in X_1(z)$ by Lemma 6.2. Thus z and x_k are colinear. The rest of the proof is by induction in $|I|$, and the arguments in the induction step are precisely those in the case $|I| = 2$. The rest of the proof follows now from Lemma 7.2. This completes the proof of Theorem 7.4. □

Case 3: $U = U(IV_n)$**, the spin factors.**

We consider now the special case of Theorem 2.4 related to the Cartan factors $U = U(IV_n)$, $4 \leq n \leq \infty$, called the spin factors (For $U = U(IV_\infty), U_p = \{0\}$). We need some notions from $[AF2]$ and $[DF]$.

Definition 7.5 (see $[AF2]$, Definition 4.4 and section 5(f)). An ordered quadruple (x_1, x_2, x_3, x_4) of normalized elements of C_p is said to satisfy the *Matrix Condition* (or condition M for short) if there exist orthogonal elements $a_1, a_2 \in C_p$ with $\|a_1\|^p + \|a_2\|^p = 1$, and a tensor product representation so that

$$x_1 = a_1 \otimes e_{1,1} + a_2 \otimes e_{1,1}; \quad x_2 = a_1 \otimes e_{1,2} + a_2 \otimes e_{2,1};$$

$$x_3 = a_1 \otimes e_{2,2} + a_2 \otimes e_{2,2}; \quad -x_4 = a_1 \otimes e_{2,1} + a_2 \otimes e_{1,2}. \tag{7.16}$$

Clearly, (7.16) can be written in matrix notation as

$$x_1 = \left(\begin{array}{cc|cc} a_1 & 0 & 0 & 0 \\ 0 & 0 & 0 & 0 \\ \hline 0 & 0 & 0 & 0 \\ 0 & 0 & 0 & 0 \end{array}\right), \quad x_2 = \left(\begin{array}{cc|cc} 0 & a_1 & 0 & 0 \\ 0 & 0 & 0 & 0 \\ \hline 0 & 0 & 0 & 0 \\ 0 & 0 & a_2 & 0 \end{array}\right),$$

$$x_3 = \left(\begin{array}{cc|cc} 0 & o & 0 & 0 \\ 0 & a_1 & 0 & 0 \\ \hline 0 & 0 & 0 & 0 \\ 0 & 0 & 0 & a_2 \end{array}\right), \quad x_4 = \left(\begin{array}{cc|cc} 0 & 0 & 0 & 0 \\ -a_1 & 0 & 0 & 0 \\ \hline 0 & 0 & 0 & -a_2 \\ 0 & 0 & 0 & 0 \end{array}\right). \tag{7.17}$$

It is easily seen that if (x_1, x_2, x_3, x_4) satisfies the condition M then the quadruple (u_1, u_2, u_3, u_4) of their partial isometries (i.e., $u_j = v(x_j)$) forms an odd quadrangle in the sense of $[DF]$. The converse is false. It is also easy to verify that if (x_1, x_2, x_3, x_4) satisfies condition M then the following conditions are satisfied:

orthogonality : $x_1 \perp x_3;\ x_2 \perp x_4$;

colinearity : $x_1 \top x_2,\ x_2 \top x_3,\ x_3 \top x_4,\ x_4 \top x_1$;

exchangeability : $2\{x_1, x_2, x_3\} = -x_4$.

It is not hard to see that these three conditions actually imply that (x_1, x_2, x_3, x_4) satisfies condition M and thus provide a characterization of condition M. Clearly, (x_1, x_2, x_3, x_4) satisfies M if and only if the same is true for any quadruple obtained by a cyclic permutation. Thus in the exchangeability condition one can apply any cyclic permutation. It is less obvious that a quadrangle can be characterized by a norm condition. In $[AF1]$ we show that if $1 \leq p < \infty, p \neq 2$, then a quadruple (x_1, x_2, x_3, x_4) of normalized elements in C_p satisfies condition M if and only if

$$\|\sum_{j=1}^{n} t_j x_j\| = \|\begin{pmatrix} t_1 & t_2 \\ -t_4 & t_3 \end{pmatrix}\| \tag{7.18}$$

for all choices of scalars t_1, t_2, t_3 and t_4 .

Definition 7.6 (see $[AF2$, Definition 5.21]).

(i) A system $(x_1, x_2, \ldots, x_n; \tilde{x}_1, \tilde{x}_2, \ldots, \tilde{x}_n), 2 \leq n < \infty$, of normalized elements of C_p is said to satisfy the *Strong Matrix Condition* (SM for short) if for any $1 \leq i, j \leq n$ with $i \neq j$, the quadruple $(x_i, x_j, \tilde{x}_i, \tilde{x}_j)$ is a quadrangle.

(ii) A system $(x_1, \ldots, x_n; \tilde{x}_1, \ldots, \tilde{x}_n; x_0), 2 \leq n < \infty$, of normalized elements of C_p is said to satisfy the *Diagonal Strong Matrix Condition* (DSM for short) if $(x_1, \ldots, x_n; \tilde{x}_1, \ldots, \tilde{x}_n)$ satisfy condition SM, x_0 governs each of the $\{x_j\}$ and $\{\tilde{x}_j\}$, and $\tilde{x}_j = -Q(v(x_0))x_j$ for every $1 \leq j \leq n$.

The following results characterize systems satisfying SM or DSM. The proof in $[AF2]$ for $p = 1$ works with slight modifications for $1 < p < \infty$ and yields the result for these values of p. The elements $x_{n,j}$ and $\tilde{x}_{n,j}$ appearing in Lemma 7.7 were introduced in $[AF2], (2.38)$ and (2.39).

Lemma 7.7 *(see [AF2, Corollary 5.26]). Let $1 \le p < \infty$. A system $(x_1, ..., x_n; \tilde{x}_1, ..., \tilde{x}_n), 2 \le n < \infty$ of normalized elements in C_p satisfies condition SM if and only if there exists orthogonal elements $a_1, a_2 \in C_p$ with $\|a_1\|^p + \|a_2\|^p = 2^{2-n}$, and there exists a tensor product representation so that for every $1 \le j \le n$:*

$$x_j = a_1 \otimes x_{n,j} + a_2 \otimes \tilde{x}_{n,j}; \quad \tilde{x}_j = a_1 \otimes \tilde{x}_{n,j} + a_2 \otimes x_{n,j}. \tag{7.19}$$

Lemma 7.8 *(see [AF2, Corollary 5.29]). Let $1 \le p < \infty$. A system $(x_1, ..., x_n; \tilde{x}_1, ..., \tilde{x}_n; x_0)$, $2 \le n < \infty$, of normalized elements of C_p satisfies condition DSM if and only if there exist an element $a \in C_p$ with $\|a\|^p = 2^{1-n}$, and a tensor product representation so that*

$$x_j = a \otimes x_{n+1,j}; \quad \tilde{x}_j = a \otimes \tilde{x}_{n+1,j}; \quad 1 \le j \le n;$$

$$x_0 = a \otimes (x_{n+1,n+1} + \tilde{x}_{n+1,n+1})/2. \tag{7.20}$$

Lemma 7.9 *Let $U = U(IV_n)$,and let $(u_j, \tilde{u}_j)_{j \in J}$ or $(u_j, \tilde{u}_j, u_0)_{j \in J}$ be a spin grid in U. Let $|n| = |J|$. In the first case define $T_1, T_2 : U \to B(H)$ by*

$$T_1(u_j) = x_{n,j}; \quad T_1(\tilde{u}_j) = \tilde{x}_{n,j}; \quad T_2(u_j) = \tilde{x}_{n,j}; \quad T_2(\tilde{u}_j) = x_{n,j}. \tag{7.21}$$

In the second case define $T_3 : U \to B(H)$ by

$$T_3(u_j) = x_{n+1,j} \quad T_3(\tilde{u}_j) = \tilde{x}_{n+1,j}; \quad T_3(u_0) = x_{n+1,n+1} + \tilde{x}_{n+1,n+1}. \tag{7.22}$$

Then T_1, T_2 and T_3 are triple monomorphisms.

The proof is contained in $[DF]$ and $[AF2]$. Indeed, it is clearly enough to show that the T_j preserve the operator norm, see section 2. The singular numbers of an element in U are computed in $[DF$, Proposition 3.6$]$, while the computation of the singular numbers of matrices in span $\{x_{n,j}, \tilde{x}_{n,j}\}_{j=1}^n$ is given in $[AF2$, Remark 2.11$]$. Comparing both methods we see that the T_j preserve the singular numbers of elements of U, and hence they preserve also the operator norm.

Theorem 7.10 *Let X be an indecomposable N-convex subspace of C_p $(1 < p < \infty, p \neq 2)$. Assume that $\dim(X) \geq 4$, and that there exist orthogonal atoms $x, \tilde{x}$ of X so that $X_1(x+\tilde{x}) = \{0\}$. Then,*

(i) $\dim X := m < \infty$.

(ii) If m is even, say $m = 2n$,there exists a system $(x_1, \ldots, x_n; \tilde{x}_1, \ldots, \tilde{x}_n)$ of atoms of X satisfying the condition SM, so that

$$X = \text{ span } \{x_{n,j}, \tilde{x}_{n,j}\}_{j=1}^{n}.$$

(iii) If m is odd, say $m = 2n+1$, then there exists atoms $x_1, .., x_n, \tilde{x}_1, \ldots, \tilde{x}_n$ of X and a normalized element $x_0 \in X$ such that the system $(x_1, \ldots, x_n; \tilde{x}_1, \ldots, \tilde{x}_n; x_0)$ satisfies condition DSM, and so that

$$X = \text{span}(\{x_j, \tilde{x}_j\}_{j=1}^{n}, x_0).$$

Moreover,

(iv) In case (ii) above there exist orthogonal elements $a_1, a_2 \in C_p$ with $\|a_1\|^p + \|a_2\|^p = 2^{2-n}$, and triple monomorphisms T_1, T_2 of $U(IV_m)$ into $B(H)$ defined by (7.21) so that in an appropriate tensor product representation

$$X = \{a_1 \otimes T_1(w) + a_2 \otimes T_2(w); w \in U(IV_m)\}; \tag{7.23}$$

(v) In case (iii) above, there exists an element $a \in C_p$ with $\|a\| = 2^{(1-n)/p}$, a triple monomorphism T_3 from $U(IV_m)$ into $B(H)$ defined by (7.22) and a tensor product representation, such that

$$X = \{a \otimes T_3(w);\ w \in U(IV_m)\}. \tag{7.24}$$

Proof: Since X is indecomposable and $X_1(x+\tilde{x}) = 0$ we have $X_0(x_1+\tilde{x}_1) = 0$. Since $x, \tilde{x}$ are atoms we have

$$X = X_2(x+\tilde{x}) = \text{ span}\{x, \tilde{x}\} + X_1(x) \cap X_1(\tilde{x}). \tag{7.25}$$

Choose a maximal colinear family $\{x_j\}_{j=1}^{n}$ of atoms of X containing $x = x_1$. From Lemmas 6.2, 6.7 and the condition $\dim X \geq 4$ it follows that this family contains at least 2 elements. Set $\tilde{x}_1 = \tilde{x}$. Using the facts that, by (7.25), $x_j \in X_1(\tilde{x}_1)$ for any $2 \leq j \leq n$, define

$$\tilde{x}_j = -Q(\,v(\tilde{x}_1 + x_1)\,)\,x_j = -2\{v(\tilde{x}_1), x_j, v(x_1)\}. \tag{7.26}$$

Then the system $(x_1, x_j, \tilde{x}_1, \tilde{x}_j)_{j \in J}$ satisfies condition M.

To show that the system $(x_j; \tilde{x}_j)_{j \in J}$ satisfies the condition SM we have to show that for any $1 \leq i, j \leq n$ with $i \neq j$ and $1 \notin \{i, j\}$, the system $(x_i, x_j, \tilde{x}_i, \tilde{x}_j)$ satisfies condition M. By applying the "Peirce calculus" (2.9) to (7.26) we see that $\tilde{x}_i \in X_1(x_j)$. Since $\tilde{x}_i$ is an atom, it follows from Lemma 6.2 that $\tilde{x}_i$ is colinear to x_j. It remains to show that

$$\tilde{x}_j = -2\{v(x_i), x_j, v(\tilde{x}_i)\}. \tag{7.27}$$

Using the Main Identity for triple products (2.1) and (7.26) we get

$$\begin{aligned} -2\{v(x_i), x_j, v(\tilde{x}_i)\} &= -2\{v(x_i), x_j, -2\{v(x_1), v(x_i), v(\tilde{x}_1)\}\} \\ &= 4\{\{v(x_i), x_j, v(x_1)\}, v(x_i), v(\tilde{x}_1)\} \\ &\quad -4\{v(x_1), \{x_j, v(x_i), v(x_i)\}, v(\tilde{x}_1)\} \\ &\quad +4\{v(x_1), v(x_i), \{v(x_i), x_j, v(\tilde{x}_1)\}\}. \end{aligned} \tag{7.28}$$

By the "Peirce calculus" (2.9)

$$\{v(x_i), x_j, v(x_1)\} \in X_2(\,v(x_i)\,) \cap X_2(\,v(x_1)\,) = \{0\}, \tag{7.29}$$

and

$$\{v(x_i), x_j, v(\tilde{x}_1)\} \in X_2(\,v(x_i)\,) \cap X_2(\tilde{x}_1) = \{0\}.$$

Since $x_j \in X_1(v(x_i))$, $\{x_j, v(x_i), v(x_i)\} = \frac{1}{2}x_j$, and so (7.26) and (7.29) imply (7.28). Thus the family $\{x_j, \tilde{x}_j\}$ satisfies the condition SM.

Let I be any finite subset of J with $i \in I$ and $|I|$=k. Then the system $\{x_i, \tilde{x}_i\}_{i \in I}$ satisfies the condition SM, and from the representation (7.19) it follows that

$$\|x_1\|_\infty = \max(\|a_1\|_\infty, \|a_2\|_\infty) \leq \max(\|a_1\|, \|a_2\| \leq 2^{(2-k)/p}.$$

Thus k is bounded above and so I is finite. We will therefore assume that $J = \{1, 2, \ldots, n\}$.

If $\bigcap_{j=1}^n X_1(x_j) = \{0\}$, then from the decomposition (7.25) and from the atomicity of $\{x_j, \tilde{x}_j\}$ we get $X = \text{span}\ \{x_j, \tilde{x}_j\}_{j=1}^n$. If $\bigcap_{j=1}^n X_1(x_j) \neq \{0\}$, let y be an atom in this N-convex subspace. If y is colinear to x_1, then by Corollary 6.6, y is an atom of X and is colinear to any $x_j, 2 \leq j \leq n$. This contradicts the maximality of the family $\{x_j\}_{j \in J}$. Thus by Lemmas 6.2 and 6.7, y governs x_1 as well as the element

$$Q(v(y))x_1 \in X_0(x_1) = X_2(\tilde{x}_1) = \mathbf{C}\tilde{x}_1.$$

Define x_0 to be a multiple of y by a complex number of absolute value 1 such that

$$Q\,(\,v(x_0)\,)\,x_1 = -\tilde{x}_1. \tag{7.30}$$

Consider the N-convex subspace $Z = \bigcap_{j=2}^n X_1(x_j)$. In this subspace x_1 is an atom, and x_0 is an atom of $Z_1(x_1)$ governing x_1 and $\tilde{x}_1$ and satisfying (7.30). Thus

$$\begin{aligned}\bigcap_{j=1}^n X_1(x_j) &= X_2(x_1 + \tilde{x}_1) \cap \bigcap_{j=1}^n X_1(x_j) = X_2(x_0) \cap X_1(x_1) \bigcap_{j=2}^n X_1(x_j) \\ &= Z_2(x_0) \cap Z_1(x_1) = \mathbf{C}x_0.\end{aligned}$$

It follows that $X = \text{span}\ (\ \{x_j, \tilde{x}_j\}_{j=1}^n, x_0\ )$.

For any $2 \leq j \leq n$, $\bigcap_{i \neq j} X_1(x_i)$ is an N-convex subspace of dimension 3, spanned by $x_j, \tilde{x}_j$ and x_0. Therefore x_0 governs x_j. From (7.30) follows the existence of an element $a \in C_p$ with $\|a\| = 1$, and a tensor product representation such that $x_1 = a \otimes e_{1,1}, \tilde{x}_1 = -a \otimes e_{2,2}$, and $x_0 = a \otimes (e_{1,2} + e_{2,1})/2^{1/p}$. Thus $Q\,(\,v(x_1 + \tilde{x}_1)\,)\,x_0 = -x_0$. By using the Main Identity as before we get for any $i \leq j \leq n$

$$\begin{aligned}Q(v(x_0))x_j &= \{v(x_0), x_j, -2\{v(x_1), v(x_0), v(\tilde{x}_1)\}\} \\ &= -2\{\{v(x_0), x_j, v(x_1)\}, v(x_0), v(\tilde{x}_1)\} \\ &\quad +2\{v(x_1)\{x_j, v(x_0), v(x_0)\}v(\tilde{x}_1)\}\end{aligned}$$

$$-2\{v(x_1), v(x_0), \{v(x_0), x_j, v(\tilde{x}_1)\}\}.$$

Using the "Peirce calculus" rule (2.9) again we see that the first and last terms belong to
$X_0(x_j) \cap X_2(x_1) = \{0\}$ and $X_0(x_j) \cap X_2(\tilde{x}_1) = \{0\}$ respectively. Thus $Q(v(x_0))x_j = -\tilde{x}_j$ and the system $(x_1, \ldots, x_n; \tilde{x}_1, \ldots, \tilde{x}_n; x_0)$ satisfies condition DSM.

So far we have established (i), (ii) and (iii) of Theorem 7.10. Statements (iv) and (v) follow from Lemmas 7.7, 7.8 and 7.9. The proof of Theorem 7.10 is complete. □

Case 4: $U = U(I_{m,n})$, $\min(m, n) \geq 2$.

The next special case of Theorem 2.4 corresponds to the Cartan factor $U(I_{m,n})$ with $\min(m, n) \geq 2$.

Theorem 7.11 *Let X be an indecomposable N-convex subspace of C_p, $1 < p < \infty$, $p \neq 2$. Suppose that there exist orthogonal atoms $x, \tilde{x}$ in X such that* $\dim X_2(x+\tilde{x}) = 4$. *Then there exist orthogonal elements $a_1, a_2 \in C_p$ with $\|a_1\|^p + \|a_2\|^p = 1$, and a tensor product representation so that*

$$X = \{a_1 \otimes w + a_2 \otimes w^T; w \in U_p(I_{n,m})\}.$$

Here w^T is the transpose of w and $2 \leq n, m \leq \infty$.

Proof: Note that the N-convex space $X_2(x + \tilde{x})$ satisfies the assumption of Theorem 7.10. Thus there are $y, \tilde{y} \in X_2(x+\tilde{x})$ such that $(x, y, \tilde{x}, \tilde{y})$ satisfies condition M. By Lemma 6.10 there exist orthogonal elements $a_1, a_2 \in C_p$ with $\|a_1\|^p + \|a_2\|^p = 1$ and a tensor product representation so that
$X_2(x + \tilde{x})$ is spanned by the elements:

$$x_{1,1} = x = a_1 \otimes e_{1,1} + a_2 \otimes e_{1,1}, \quad x_{1,2} = y = a_1 \otimes e_{1,2} + a_2 \otimes e_{2,1},$$

$$x_{2,1} = \tilde{y} = a_1 \otimes e_{2,1} + a_2 \otimes e_{1,2}, \quad x_{2,2} = -\tilde{y} = a_1 \otimes e_{2,2} + a_2 \otimes e_{2,2}. \tag{7.31}$$

That is, $(x_{1,1}, x_{1,2}, x_{2,2}, -x_{2,1})$ satisfies condition M. Since $X_1(x_{1,2}) \cap X_1(x_{2,1}) = \text{span}\{x_{1,1}, x_{2,2}\}$ we have $X_1(x_{1,1}) \cap X_1(x_{1,2}) \cap X_1(x_{2,1}) = \{0\}$. Thus $X_1(x_{1,1})$ decomposes as:

$$X_1(x_{1,1}) = [X_1(x_{1,1}) \cap X_0(x_{1,2})] + [X_1(x_{1,1}) \cap X_0(x_{2,1})] := Z_1 + Z_2.$$

We claim that Z_1 is orthogonal to Z_2. Obviously, $x_{2,1}$ is orthogonal to Z_2, therefore in order to prove that Z_2 is orthogonal to Z_1 it is enough to prove that Z_2 is orthogonal to any $y \in Z_1 \cap X_1(x_{2,1})$ of norm 1. By Lemma 6.2 and Corollary 6.6 y is colinear to $x_{2,1}$ and by Lemma 6.5 there is an isometry Φ of the N-convex subspace $X_1(x_{1,1})$ exchanging y and $x_{2,1}$. By Lemma 6.5(iv), $\Phi x_{1,2} = x_{1,2}$ and thus $\Phi(Z_2) = Z_2$. Using the fact that $x_{2,1} = \Phi(y)$ is orthogonal to $\Phi(Z_2) = Z_2$ and the fact that any isometry preserves orthogonality, we obtain that y is orthogonal to Z_2. Thus Z_1 is orthogonal to Z_2.

Since the rank of $X_1(x_{1,1})$ is 2 it follows that Z_1, Z_2 are N-convex subspaces of rank 1 and by Theorem 7.4 there are two colinear families of atoms of X : $\{x_{1,j}\}_{j=2}^m$ and $\{x_{i1}\}_{i=2}^n$ such that $Z_1 = \overline{\text{span}}\{x_{i1}\}_{i=2}^n$, $Z_2 = \overline{\text{span}}\{X_{1j}\}_{j=2}^m$. Applying Proposition 6.3, the tensor representation (7.31) can be extended is such a way that

$$x_{i,1} = a_1 \otimes e_{i,1} + a_2 \otimes e_{1,i}, \quad x_{1,j} = a_1 \otimes e_{1,j} + a_2 \otimes e_{j,1} \tag{7.32}$$

for all $1 \leq i \leq n$ and $1 \leq j \leq m$. By Corollary 4.3 the elements $x_{i,j}$ defined by

$$x_{i,j} = Q(v(x_{i,1} + x_{1,j}))x_{1,1} = a_1 \otimes e_{i,j} + a_2 \otimes e_{j,i}$$

for $1 \leq i \leq n$, $1 \leq j \leq m$ belong to X. Set $U = \overline{\text{span}}^{w^*}\{e_{i,j}\} \,; 1 \leq i \leq m, 1 \leq j \leq n\}$ where the closure is taken in the w^*-topology. Then $U = U(I_{n,m})$ and thus it is enough to show that

$$X = \overline{\text{span}}\{x_{i,j}\};\ 1 \leq i \leq n, 1 \leq j \leq m\} \tag{7.33}$$

where the closure is taken in the norm of C_p. Fix $2 \leq i_0 \leq n$. Then by Lemma 6.5 there exists a surjective isometry Φ of C_p which interchanges $x_{1,1}$ and $x_{i_0,1}$. By the proof of Lemma 6.5 Φ maps $Y := \overline{\text{span}}\{x_{i,j};\ 1 \le i \le n, 1 \le j \le m\}$ onto itself and maps $X_1(x_{1,1})$ onto $X_1(x_{i_0,1})$, by the maximality of the families $\{x_{1,j}\}_{j=1}^m$ and $\{x_{i,1}\}_{i=1}^n$. Thus $X_1(x_{i,1}) \subseteq Y$ for every $1 \le i \le n$. Similarly $X_1(x_{1,j}) \subseteq Y$ for every $1 \le j \le m$. Thus

$$X = Y + (\bigcap_{1 \le i \le n} X_0(x_{i,1})) \cap (\bigcap_{1 \le j \le m} X_0(x_{1,j})).$$

Since X is indecomposable the right summand is zero. Thus $X = Y$ and (7.33) holds. □

Case 5: $U = U(II_n)$, the anti-symmetric n by n matrices.

We turn now to the special case of Theorem 2.4 corresponding to $U = U(II_n)$. The proof is based on the following lemma, analogous to Proposition 4.9 in [$AF2$].

Lemma 7.12 *Let X be an N-convex subspace of C_p $(1 < p < \infty,\ p \ne 2)$. Assume that there exist orthogonal atoms $x, \tilde{x}$ in X such that* $\dim X_2(x+\tilde{x}) > 4$ *and $X_1(x+\tilde{x})$ is not the zero space. Then* $\dim X_2(x+\tilde{x}) = 6$. *Moreover, there exist atoms $y, \tilde{y}, w, \tilde{w}$ of X such that $X_2(x+\tilde{x}) = \text{span}\{x, \tilde{x}, y, \tilde{y}, w, \tilde{w}\}$, an operator $a \in C_p$ with $2\|a\|^p = 1$, and a tensor product representation such that*

$$x = a \otimes (e_{1,2} - e_{2,1}); \quad y = a \otimes (e_{1,3} - e_{3,1}); \quad w = a \otimes (e_{1,4} - e_{4,1});$$

$$\tilde{x} = a \otimes (e_{3,4} - e_{4,3}); \quad \tilde{y} = a \otimes (e_{2,4} - e_{4,2}); \quad \tilde{w} = a \otimes (e_{2,3} - e_{3,2}).$$

Proof. Since $\dim X_2(x + \tilde{x}) > 4$ there are atoms $y, \tilde{y}$ of $X_2(x + \tilde{x})$ such that the system $(x, y, \tilde{x}, \tilde{y})$ satisfies condition M. It is easy to verify (see also (2.1) of [DF]) that $X_1(x + \tilde{x})$ can be decomposed as:

$$X_1(x + \tilde{x}) = [X_1(x) \cap X_0(\tilde{x})] + [X_0(x) \cap X_1(\tilde{x})].$$

Similarly

$$X_1(y + \tilde{y}) = X_1(x + \tilde{x}) = [X_1(y) \cap X_0(\tilde{y})] + [X_0(y) \cap X_1(\tilde{y})].$$

Using the fact that $X_1(x+\tilde{x}) \neq \{0\}$ we may assume without loss of generality that there is a normalized element z in $X_0(\tilde{x}) \cap X_1(x)$ which is orthogonal to $\tilde{y}$. By Corollary 6.6 that z is colinear to x, and y and it is an atom of X. By (2.9) the involution

$$Q(y+\tilde{y}) = 2\{y, \cdot, \tilde{y}\}$$

on $X_2(x+\tilde{x})$ interchanges the subspaces

$$Z_1 = X_1(z) \cap X_1(x) \cap X_1(\tilde{x}) \cap X_1(y)$$

and

$$Z_2 = X_0(z) \cap X_1(x) \cap X_1(\tilde{x}) \cap X_1(y).$$

Thus $\dim Z_1 = \dim Z_2$, $\dim X_2(x+\tilde{x}) = 4 + 2\dim Z_1$ is even, and $X_1(x+\tilde{x}) \neq \{0\}$. Let w be a normalized element in Z_1 and let $\tilde{w} = -Q(y+\tilde{y})w \in Z_2$. Since w and $\tilde{w}$ are orthogonal elements in $X_1(x)$ they must be atoms of X_1 and both are colinear to $x, \tilde{x}, y$ and $\tilde{y}$.

Next, we show that $\dim Z_1 = 1$. Using the relations between the element $x, \tilde{x}, y, \tilde{y}, w$ and $\tilde{w}$ we can decompose each of them into two orthogonal parts ($x = x_1 + x_2, \tilde{x} = \tilde{x}_1 + \tilde{x}_2$, etc.) such that in an appropriate matrix representation they are described schematically as follows:

$$\left(\begin{array}{cccc|c} 0 & x_1 & y_1 & w_1 & z_1 \\ x_2 & 0 & \tilde{w}_2 & \tilde{y}_2 & * \\ y_2 & \tilde{w}_1 & 0 & \tilde{x}_2 & * \\ w_2 & \tilde{y}_1 & \tilde{x}_1 & 0 & * \\ \hline z_2 & * & * & * & * \end{array}\right) \tag{7.34}$$

with

$$|x_1^*| = |y_1^*| = w_1^*| = |z_1^*|, \quad |x_2| = |y_2| = |w_2| = |z_2| \tag{7.35}$$

and similarly for all other "rows" and "columns" in (7.34). From the definition of Z_1 any element $b \in Z_1$ satisfies:

$$b = P_1(z)P_1(x)P_1(\tilde{x})P_1(y)b. \tag{7.36}$$

Hence,

$$\begin{aligned} b &= \ell(x_1)br(\tilde{x}_2) + \ell(\tilde{x}_1)br(x_2) \\ &= [\ell(w_1) + \ell(w_2)]b[r(w_1) + r(w_2)] = P_2(w)b = \lambda w \end{aligned}$$

for some constant λ. Thus $\dim Z_1 = 1$ and $\dim X_2(x + \tilde{x}) = 6$.

It follows that the spanning system $(x, y, w; \tilde{x}, \tilde{y}, \tilde{w})$ of $X_2(x+\tilde{x})$ satisfies condition SM, and so it is described by Theorem 7.10 with $m = 6$. Notice that by $[AF2$, p.25$]$ the matrices occuring in (7.19) correspond to the elementary anti-symmetric 4×4 matrices. Also, from the matrix representation (7.34) it follows that in the tensor product representation (7.19) one of a_1, a_2 vanishes. Without loss of generality we assume that $a_2 = 0$. Thus (7.19) and (7.34) give a tensor product representation such that

$$x = a \otimes (e_{1,2} - e_{2,1}); \quad y = a \otimes (e_{1,3} - e_{3,1}); \quad w = a \otimes (e_{1,4} - e_{4,1}); \tag{7.37}$$

$$\tilde{x} = a \otimes (e_{3,4} - e_{4,3}); \quad \tilde{y} = a \otimes (e_{2,4} - e_{4,2}); \quad \tilde{w} = a \otimes (e_{2,3} - e_{3,2}).\square$$

Remark 7.13 The proof that $\dim Z_1 = 1$ in Lemma 7.12 is different in nature than all previous proofs given in this section. While all other proofs use only abstract "Peirce calculus" (2.9) and therefore are valid also in the context of the exceptional Cartan factors, the proof of "$\dim Z_1 = 1$" uses the fact that $C_p \subset B(H)$ and so the Peirce projection are given in terms of the support projections of the corresponding element. In the general JB^*-triple setup, $\dim Z_1$ can be 1,2 or 3, where the two latter cases correspond to the exceptional Cartan factors of type V and VI of dimensions 16 and 27 respectively, see $[DF]$.

Theorem 7.14 *Let X be an indecomposable N-convex subspace of C_p $1 < p < \infty$, $p \neq 2$.* Assume that there exist *orthogonal atoms $x, \tilde{x}$ in X such that* $\dim X_2(x+\tilde{x}) > 4$ *and $X_1(x + \tilde{x}) \neq \{0\}$. Then there exist a system $\{x_{i,j}\}_{1 \leq i < j \leq n}$ of atoms of X with $4 < n < \infty$, an operator $a \in C_p$ with $2\|a\|^p = 1$, and a tensor product* *representation such that*

$$x_{i,j} = a \otimes (e_{i,j} - e_{j,i}), \qquad 1 \leq i < j \leq n \tag{7.38}$$

and

$$X = \overline{\text{span}}\{x_{i,j}\}_{1 \leq i < j \leq n}. \tag{7.39}$$

Moreover, we can arrange that $x_{1,2} = x$ *and* $x_{3,4} = \tilde{x}$.

Denote $U = \overline{\text{span}}^{w^*}(e_{i,j} - e_{j,i})_{1 \leq i < j \leq n}$, *where the closure is in the* w^**-topology. Then* $U = U(II_n)$ *and*

$$X = \{a \otimes w;\ w \in U_p\}. \tag{7.40}$$

Proof: Since X satisfies the conditions of Lemma 7.12 there is an element $a \in C_p$, $2\|a\|^p = 1$ and a tensor product representation such that (7.38) holds for atoms $x_{i,j}$ of X, $1 \leq i < j \leq 4$ and $x = x_{1,2}$, $\tilde{x} = x_{3,4}$. The N-convex subspace $X_1(x_{1,2})$ is of rank 2 and satisfies the conditions of Theorem 7.11. Thus there is an extension of the tensor product representation such that all the $x_{i,j}$ defined by (7.38) are atoms of x, for i=1,2 and $3 \leq j \leq n$.
Moreover,

$$X_1(x_{1,2}) = \overline{\text{span}}\{x_{i,j} : \ i = 1, 2, 3 \leq j \leq n\}. \tag{7.41}$$

For any $3 \leq i, j \leq n$ with $i < j$ define

$$x_{i,j} = Q(x_{1,i} + x_{2,j})x_{1,2}.$$

Then, obviously, $x_{i,j}$ satisfy (7.38) and are atoms of X. Letting $Y = \overline{\text{span}}\{x_{i,j},\ 1 \leq i < j \leq n\}$, by (7.41) we have $X_1(x_{1,2}) \subset$ Y. As in the proof of Theorem 7.11, we apply Lemma 6.4 to obtain that $X_1(x_{i,j}) \subset Y$ for any $2 \leq j$. Thus $X = Y + \bigcap_{1 \leq i < j \leq n} X_0(x_{i,j})$. Since X is indecomposable we get $X = Y$ and (7.39) follows. The other statements in Theorem 7.14 are now obvious. □

We are now able to prove Theorem 2.4

Proof of Theorem 2.4

Choose an atom x in X. If $X_1(x) = \{0\}$, then $X_0(x) = \{0\}$ as well since X is indecomposable. Thus $\dim X = 1$ and the theorem is obvious. Otherwise $X_1(x) \neq \{0\}$. By Corollary 6.11, rank $[X_1(x)]$ is 1 or 2. If rank $[X_1(x)] = 1$ we choose an atom y of $X_1(x)$. By Lemma 6.2, either y is colinear to x and then the result follows from Theorem 7.4, or $x \in X_2(y)$ and the result follows from Theorem 7.1. If rank $[X_1(x)] = 2$, there is a non-zero element $y \in X_1(x)$, which is not an atom of $X_1(x)$. Then by lemma 6.10 there are atoms $z, \tilde{z}$ and $\tilde{x}$ of X so that $z, \tilde{z} \in X_2(x + \tilde{x})$ and $(x, z, \tilde{z}, \tilde{x})$ satisfy condition M. If $X_1(x + \tilde{x}) = \{0\}$ then the result follows from Theorem 7.10, and if $\dim X_2(x + \tilde{x}) = 4$ from Theorem 7.11. The remaining case is covered by Theorem 7.14.

8. Conclusion of the proof of the Main Theorem and applications

In this section we conclude the proof of the Main Theorem, and obtain more results on the structure of ranges of contractive projections in $C_p \, (1 < p < \infty, p \neq 2)$.

The equivalence of statements (1) and (3) in the Main Theorem was proved in Corollary 1.3. Proposition 2.2 reduces the Main Theorem to indecomposable subspaces, and in the rest of the work we consider mainly indecomposable subspaces. Proposition 2.3 is the implication (2) $\Rightarrow$ (1) in the Main Theorem, and Theorem 2.4 is the implication (1) $\Rightarrow$ (2). Theorem 2.4 is proved in Section 7 after the long preparation in Sections 4, 5 and 6. The equivalence (1) $\Leftrightarrow$ (2) is expressed in Corollary 2.5. It remains to prove the equivalence (1) $\Leftrightarrow$ (4), and this is our first goal in this section.

We begin with the implication (1) $\Rightarrow$ (4) which is an easy Corollary of Theorem 2.4.

Corollary 8.1. *Let X be a subspace of $C_p, 1 < p < \infty, p \neq 2$, and assume that X*

is the range of a contractive projection from C_p. Let $V := \overline{span}^{w^*}\{v(x); x \in X\}$, where the closure is taken in the w^*-topology. Then V is a atomic JCW^*-subtriple of $B(H)$ and X is a module over V, namely

$$\{V, V, X\} \subset X \quad \textit{and} \quad \{V, X, V\} \subset X. \tag{8.1}$$

Proof: By Proposition 2.2, it is enough to consider the case where X is indecomposable. We apply Theorem 2.4, and assume that X is given by (2.27). Using the facts that $v(a \otimes b) = v(a) \otimes v(b)$ and $v(x + y) = v(x) + v(y)$ for orthogonal elements x and y, and that $v(T(w)) = T(v(w))$ for a triple monomorphism T we get that

$$V = \{\sum_{j \in J} v(a_j) \otimes T_j(w); w \in U\} \tag{8.2}$$

and that the map $\tilde{T} : U \to V$ defined by

$$\tilde{T}(w) := \sum_{j \in J} v(a_j) \otimes T_j(w) \tag{8.3}$$

is a triple isomorphism. Thus V is an atomic JCW^*-subtriple of $B(H)$. Next, U_p is a module over U, namely for $w_1, w_2 \in U$ and $w_3 \in U_p$,

$$\{\tilde{T}(w_1), \tilde{T}(w_2), T(w_3)\} = T\{w_1, w_2, w_3\}, \tag{8.4}$$

and

$$\{\tilde{T}(w_1), T(w_3), \tilde{T}(w_2)\} = T\{w_1, w_3, w_2\}. \tag{8.5}$$

Both formulas are easy consequences of the definitions (2.26), (8.3) and the fact that the T_j are triple monomorphisms. This proves the module property (8.1) of X. □

The implication (4) ⇒ (2) in the Main Theorem is proved in the following Proposition.

Proposition 8.2. *Let X be a subspace of $C_p, 1 < p < \infty$, and define $V := \overline{span}^{w^*}\{v(x); x \in X\}$, the closure being taken in the w^*-topology. Assume that*

V is an atomic JCW^*-triple and that X is a module over V, namely that (8.1) holds. Then X is the ℓ_p-sum of subspaces, each of which is triple-isometric to the C_p-ideal of a classical Cartan factor.

The following lemma is the key to Proposition 8.2.

Lemma 8.3. *Let X and V be as in Proposition 8.2 and let $v \in V$ be a minimal tripotent. Then there exists an atom x of X so that $v = v(x)$.*

Proof: Since the Peirce projections $P_k(v)$ $(k = 0,1,2)$ are polynomials in $D(v) = \{v, v, \cdot\}$, we get by (8.1) that $P_k(v)$ maps X into X $(k = 0,1,2)$. Thus $X_k(v) = P_k(v)X$ for $k = 0,1,2$ and $X = X_2(v) + X_1(v) + X_0(v)$. We claim that $X_2(v) \neq 0$. If $X_2(v) = \{0\}$ and also $X_1(v) = \{0\}$ then $X = X_0(v)$ and necessarily $V = V_0(v)$, contradicting the fact that $v \in V$. Thus $X_1(v) \neq 0$.

Fix a normalized element $y \in X_1(v)$ and let $u = v(y)$ be the tripotent from the polar decomposition of y. By [DF, Proposition 2.1], either u is colinear to v $(u \top v)$ or u governs v $(u \vdash v)$, see also Section 6 for these notions. If $u \top v$ let $w = (u+v)/\sqrt{2}$. As $w \in V$, $S(w) := e^{2\pi i D(w)}$ maps X into itelf by (8.1). If $x = S(w)y$, then $x \in X$ and $S(w)u = v$. Since $S(w)$ is a triple automorphism we get $v = v(x)$ and in particular, $X_2(v) \neq \{0\}$.

If $u \vdash v$, define $w = (v + u + Q(u)v)/2$. As $w, v \in V$ and $y \in X$ we get by (8.1) that $P_2(v)Q(w)y$ is a non zero element of $X_2(v)$. Thus, indeed $X_2(v) \neq \{0\}$.

We claim next that $\dim X_2(v) = 1$. If not, let x_1, x_2 be linearly indepenent normalized elements of $X_2(v)$. As $v(x_j) \in V_2(v)$ and v is a minimal tripotent in V (hence an atom of V, see Section 2), we have $v(x_j) = \lambda_j v$ with $|\lambda_j| = 1, j = 1,2$. By replacing x_j by $\bar{\lambda}_j x_j$, we can assume without loss of generality that $v(x_j) = v, j = 1,2$. It follows that $x_j = Q(v)x_j$, *i.e.*, both x_1 and x_2 are self-adjoint with respect to v. If $x_1 \neq x_2$, there are vectors $\xi_1, \xi_2 \in H$ and a real number t so that

$$((x_1 + tx_2)\xi_1, v\xi_1) > 0 > ((x_1 + tx_2)\xi_2, v\xi_2).$$

But this implies that $v_1 = v(x_1 + tx_2)$ is not proportional to v. Using the fact that $v_1 \in V_2(v)$ is self-adjoint with respect to v and v is an atom of V, we see that v_1 must be a real multiple of v. This contradiction proves that $x_1 = x_2$ and so $\dim X_2(v) = 1$. As $X_2(v) = X_2(x), x$ is an atom of X satisfying $v = v_2(x)$. □

Proof of Proposition 8.2. By Proposition 2.2, X is the orthogonal direct sum of indecomposable subspaces $\{X_k\}$ (thus $X = (\sum_k \oplus X_k)_p$). Let $V_k = \overline{\text{span}}^{w^*}\{v(x);\ x \in X_k\}$. Then V_k is an indecomposable ideal in V and (8.1) holds for X_k, V_k instead of X and V. Without loss of generality we assume therefore that X and V are indecomposable.

Being indecomposable, V is triple-isomorphic to a classical Cartan factor U. Fix a colinear-orthogonal-governing (cog) grid of tripotents in V whose linear span is w^*-dense in V, and fix a minimal tripotent v from the grid. By Lemma 8.3 there exists an atom $x \in X$ so that $v = v(x)$. We construct now, by the module property (8.1), a cog-grid in X, consisting of elements whose supporting tripotents are the elements of the above grid of V. Indeed, if u is a minimal tripotent from the grid which is colinear to v, let $w = (v + u)/\sqrt{2}$ and let $y = e^{2\pi i D(w)}x$. Then $y \in X, v(y) = u$ and y is an atom of X. These properties are easily derived from Definition 6.1 and the fact that $e^{2\pi i D(w)}$ is a triple-automorphism. If $(v, u, \tilde{v})$ is a trangle from the grid of V, define $y = 2^{-1/p}D(\tilde{v}, u)x$ and $\tilde{x} = Q(u)x$. Then $(x, y, \tilde{x})$ is a trangle of elements of X, $v(y) = u, v(\tilde{x}) = v$. Moreover, $\tilde{x}$ is an atom of X and y is an atom of $X_1(x) \cap X_1(\tilde{x})$. These facts for $\tilde{x}$ follow from Definition 6.1 and the fact that $Q(u)$ is an involution on $X_2(u)$. The statements concerning y can be obtained also along these lines. Alternatively—use Lemma 8.3 with $X_2(u) \cap X_1(x)$ and $V_2(u) \cap V_1(v)$ instead of X and V.

Applying the proofs of Section 7 to the cog-grid just constructed in X, we conclude that X has the form (2.27) and that it is triple-isometric to U_p. □

Let $\{U_j\}_{j \in J}$ be an orthogonal family of classical Cartan subfactors of $B(H)$ and let $U = \left(\sum_{j\in J} U_j\right)_\infty = \{\sum_{j\in J} x_j; x_j \in U_j, \sup_{j\in J} \|x_j\|_\infty < \infty\}$. The *$C_p$-ideal of* U is defined by $U_p = U \cap C_p$. Obviously, $U_p = \left(\sum_{j\in J}(U_j)_p\right)_p = \{\sum_{j\in J} x_j; x_j \in (U_j)_p, \sum_{j\in J} \|x_j\|_p^p < \infty\}$. If $T : U \to B(H)$ is a faithful triple representation and $V = T(U)$ is the corresponding JCW^*-triple, there are in general many ways to associate to V a "C_p-space". To illustrate this, write T in the general form

$$T(\sum_{j\in J} x_j) = \sum_{j\in J}\sum_{i\in I_j} T_{i,j}(x_j), \quad x_j \in U_j, \tag{8.6}$$

with $T_{i,j} : U_j \to B(H)$ irreducible, faithful triple representations, and the ranges of the entire family $\{T_{i,j}; i \in I_j\ j \in J\}$ are pairwise orthogonal. Consider positive weights $\mu = \{\mu_{i,j}\}_{j\in J, i\in I_j}$, let $\mu_j = \{\mu_{i,j};\ i \in I_j\}$ and assume that $\|\mu_j\|_p := (\sum_{i\in I_j} \mu_{i,j}^p)^{1/p} = 1$ for all $j \in J$. Define an isometry T_μ from U_p into C_p by

$$T_\mu(\sum_{i\in J} x_j) = \sum_{j\in J}\sum_{i\in I_j} \mu_{i,j} T_{i,j} x_j, \quad \sum_{j\in J} x_j \in U_p. \tag{8.7}$$

We denote $V_p(\mu) = T_\mu(U_p)$ and call $V_p(\mu)$ the *C_p-space associated with V and μ.*

Definition 8.4. *$\mathcal{C}_p$ is the class of all subspaces of C_p of the form $V_p(\mu)$ for JCW^*-subtriples V and appropriate family μ of weights.*

Corollary 8.5. (a) *The family $\mathcal{C}_p$ coincides with the N-convex subspaces of C_p, i.e. with the ranges of contractive projections on C_p;*

(b) $\mathcal{C}_p$ is closed under contractive projections.

Proof: To establish (a), it is enough by Theorem 2.4 to show that the family of operators T_μ in (8.6) corresponding to a set J of cardinality 1 (i.e. U is a classical Cartan factor) coincides with the family of operators of the form (2.26) considered in Proposition 2.3. This follows easily by writing the elements $a_j \in C_p$ in (2.26) in a diagonal form ($a_j = \sum_{k\in I_j} s_{k,j} v_{k,j}$, with $s_{k,j} > 0$ and $\{v_{k,j}\}_{k\in I_j, j\in J}$ orthogonal tripotents of rank 1). Next, if $X \in \mathcal{C}_p$ is the range of a contractive projection P_1 on C_p and P_2 is a contractive projection on X with range Y, then P_2P_1 is a contractive projection on C_p with range Y. Hence $Y \in \mathcal{C}_p$. This proves (b). □

Remark 8.6. Corollary 8.5 is the C_p-analogue of the facts that *the category of JB^*-triples as well as the subcategory of JC^*-triples are closed under contractive projections.* See [K] for the general JB^*-triple case, and [FR1] for the JC^*-triple case (see also [ES] for the special case of positive contractive projections on a JB^*-algebra).

The *rank* of a subspace X of C_p is the maximal cardinality of a subset of pairwise orthogonal non-zero elements of X. Clearly, the rank of $V_p(\mu)$ is equal to the rank of V as a JCW^*-triple (the cardinality of a maximal family of orthogonal tripotents).

Definition 8.7. $\mathcal{C}_p^{(1)}$ is the class of all spaces in $\mathcal{C}_p$ having no indecomposable direct summand of rank 1.

Proposition 8.8. *Let $1 \leq p < \infty, p \neq 2$. Then $\mathcal{C}_p^{(1)}$ is closed under isometries.*

This is a Corollary of [AF1] and [AF2]. Since isometries preserve orthogonal direct sums it is enough to show that if U is a classical Cartan factor of *rank* ≥ 2 then an isometric image of U_p in C_p belongs tto $\mathcal{C}_p^{(1)}$. The case $U = U(I_{n,m}), \min\{n,m\} \geq 2$, is settled in [AF1]. In [AF2, Chapter 6] we prove the corresponding result for $p = 1$ and general U with rank≥ 2. The proofs in [AF2, Chapter 6], combined with [AF1] give the desired result for $1 < p < \infty$, $p \neq 2$, and U general.

Remark 8.9. (a) Proposition 8.8 is not true in $\mathcal{C}_p$, as the Hilbert spaces $U_p(I_{1,n})$ can (for some values of p and n) be embedded isometrically into C_p (and even into ℓ_p, see [Mi]) not via the triple isometries T_μ.
(b) It is an open problem whether the analogue of Proposition 8.8 holds for $p = \infty$.
(c) $\mathcal{C}_p^{(1)}$ is not closed under contractive projections.

In Proposition 2.3 we considered a special case of triple-isometries. We define now the triple isometries in general. Let $\{U_j\}_{i\in J}, U, \{T_{i,j}; i \in I_j, j \in J\}, T, V, \mu = \{\mu_{i,j}; i \in I_j, j \in J\}, T_\mu$ and $V_p(\mu)$ be as in the discussion preceeding Definition 8.4.

Construct in the same way S_ν and $W_p(\nu)$ from the data $\{S_{k,j}; k \in K_j, j \in J\}, S, W$, and $\nu = \{\nu_{k,j}; k \in K_j, j \in J\}$. Then $S_\nu T_\mu^{-1} : V_p(\mu) \to W_p(\nu)$ is a surjective isometry. We call isometries of this type *triple isometries.* The following is obvious.

Proposition 8.10. Let $V_p(\mu)$ and $W_p(\mu)$ be members of $\mathcal{C}_p$. Then

(a) *$V_p(\mu)$ and $W_p(\nu)$ are triple isometric if and only if the underlying JCW^*-triples V, W are triple-isometric.*

(b) *If $V_p(\mu)$ and $W_p(\nu)$ belong to the subclass $\mathcal{C}_p^{(1)}$, then they are isometric if and only if they are triple-isometric.*

Moreover,

(c) *The superposition of two triple-isometries is a triple-isometry.*

Let $X, Y \in \mathcal{C}_p$. A *morphism* from X into Y is an operator of the form JAP, where P is a contractive projection on X, A is a triple isometry from $P(X)$ onto a subspace Y_1 of Y, and $J : Y_1 \to Y$ is the inclusion map. The composition of two morphisms, say JAP from X into Y and J_1, A_1, P_1, from Y into Z, is defined only if $P_1(APX) \subset AP(X)$, by the usual composition rule.

Proposition 8.11. **(a)** *The composition of two morphisms in $\mathcal{C}_p$, whenever defined, is a morphism in $\mathcal{C}_p$;*

(b) *The class $\mathcal{C}_p^*$ of all duals of spaces from $\mathcal{C}_p$ coincides with $\mathcal{C}_q\,(\frac{1}{p} + \frac{1}{q} = 1)$. Similarly, $\mathcal{C}_p^{(1)^*} = \mathcal{C}_q^{(1)}$.*

(c) *The adjoint of a morphism in $\mathcal{C}_p$ is a morphism in $\mathcal{C}_q$.*

(d) *$\mathcal{C}_p$ is closed under orthogonal direct sums.*

Proof: The proof of (a) is straight forward. (b) follows from the fact that $V_p(\mu)^* = V_q(\mu^{p-1}), 1 \leq p < \infty, \frac{1}{p} + \frac{1}{q} = 1$, where $(\mu^{p-1})_{i,j} := \mu_{i,j}^{p-1}$. To prove (c), observe first that if $T_\mu : U_p \to V_p(\mu)$ is defined via (8.7) then $T_\mu^* : V_q(\mu^{p-1}) \to U_q$ is simply $T_\mu^* = T_{\mu^{p-1}}^{-1}$. Suppose now that $JAP : X \to Y$ is a morphism in $\mathcal{C}_p$, where P is a contractive projection from X onto a subspace $V_p(\mu), W_p(\nu)$ is a subspace of Y, $A = S_\nu T_\mu^{-1} : V_p(\mu) \to W_p(\nu)$ a triple isometry and $J : W_p(\nu) \to Y$ the inclusion map. Then $(JAP)^* = P^* A^* J^* : Y^* \to X^*$, where J^* is a contractive projection onto $W_p(\nu)^* = W_q(\nu^{p-1}), V_p(\mu)^* = V_q(\mu^{p-1}), A^* = (T_\mu^{-1})^* S_\nu^* = T_{\mu^{p-1}} S_{\nu^{p-1}}^{-1}$, and P^* the inclusion map of $V_q(\mu^{p-1})$ into X^*. Thus $P^* A^* J^*$ is a morphism of $\mathcal{C}_q$. The statement (d) is obvious. □

It is important to view $\mathcal{C}_p$ also from the point of view to the *category of* JC^**-modules.*

Definition 8.12. Let V be an atomic JCW^*-subtriple of $B(H)$. A *triple-module* (or, a module, for short) over V is a linear subspace X of V which is a Banach space with respect to a norm $\|\cdot\|_X$ which in general is different from the operator norm, so that the following are satisfied:

(a) $\{V, V, X\} \subset X, \{V, X, V\} \subset X$.

(b) For every $v_1, v_2 \in V$ and $x \in X$,

$$\|\{v_1, v_2, x\}\|_X \leq \|v_1\|_\infty \|v_2\|_\infty \|x\|_X, \quad \|\{v_1, x, v_2\}\|_X \leq \|v_1\|_\infty \|x\|_X \|v_2\|_\infty, \tag{8.8}$$

where $\|\cdot\|_\infty$ is the $B(H)$-norm;

(c) For every $v \in V$, the operator $D(v) = \{v, v, \cdot\}$, considered as an operator on X, is Hermitian and has a non-negative spectrum. Moreover $\|D(v)\|_{B(X)} = \|v\|_\infty^2$.

(d) For every $v \in V$, the operator $Q(v) = \{v, \cdot, v\}$ on X satisfies $\|Q(v)\|_{B(X)} = \|v\|_\infty^2$

Proposition 8.13. *Let $X \in \mathcal{C}_p$ and let $V = \overline{span}^{w^*}\{v(x); x \in X\}$, closure in the w^*-topology. Then X is a module over V.*

We omit the straight-forward proof.

Remark 8.14 (a) If C_E is any symmetric normed ideal (see [GK1], [A3]) and U a classical Cartan factor, define $U_E = U \cap C_E$. Then U_E is a module over U. Similar statement is true with respect to orthogonal direct sums of classical Cartan factors.

(b) In Definition 8.13 and above we restrict ourselves to atomic JCW^*-triples. One can define modules over a much larger class of JC^*-triple (consider for instance $L^p(M, \tau)$, where M is a semifinite von-Neumann algebra and τ is a trace on M).

The *objects* in the category of modules over atomic JCW^*-triples are pairs (V, X), where V is an atomic JCW^*-triple and X is a module over V. A *morphism from* (V, X) *into* (W, Y) in this category is a pair $T = (T_1, T_2)$ where $T_1 : V \to W$ and $T_2 : X \to Y$ are bounded operators so that for every $v, u \in V$ and $x \in X$

$$
\begin{aligned}
T_2\{v, u, x\} &= \{T_1 v, T_1 u, T_2 x\} \\
T_2\{v, x, u\} &= \{T_1 v, T_2 x, T_1 u\}.
\end{aligned}
$$

It is easy to verify that the composition of two morphisms is again a morphism. Thus the inverse of a morphism, whenever it exists is a morphism. Thus we get a category in the usual sense.

Example 8.15. Let $a \in C_p$, $\|a\|_p = 1$. Define $T_1 : B(H) \to B(H)$ and $T_2 : C_p \to C_p$ by $T_1 x = v(a) \otimes x$, $T_2 y = a \otimes y$. Then $(v(a) \otimes B(H), a \otimes C_p)$ is an object, namely $a \otimes C_p$ is a module over $v(a) \otimes B(H)$, and $T = (T_1, T_2)$ is a morphism from $(B(H), C_p)$ onto $(v(a) \otimes B(H), a \otimes C_p)$.

Example 8.15 motivates the following result.

Proposition 8.16. *Every triple isometry between members of $\mathcal{C}_p$ determines uniquely an isometry in the category of modules. Precisely, if $S_\nu T_\mu^{-1} : V_p(\mu) \to W_p(\nu)$ is a* *triple isometry, then* $(ST^{-1}, S_\nu T_\mu^{-1}) : (V, V_p(\mu)) \to (W, W_p(\nu))$ *is a morphism in the category of modules.*

Proof: Let $\{U_j\}_{j\in J}, U, \{T_{i,j}; i \in I_j, j \in J\}, T, V, \mu = \{\mu_{i,j}; i \in I_j, j \in J\}, T_\mu$ and $V_p(\mu)$ be as in the discussion preceeding Definition 8.4. Let $a = \sum_{j\in J} a_j, b = \sum_{j\in J} b_j \in U$ (with $a_j, b_j \in U_j$) and let $x = \sum_{j\in J} x_j \in U_p$, with $x_j \in (U_j)_p$. Then

$$\begin{aligned} T_\mu\{a,b,x\} &= T_\mu \sum_{j\in J}\{a_j, b_j, x_j\} \\ &= \sum_{j\in J}\sum_{i\in I_j} \mu_{i,j} \quad T_{i,j}\{a_j, b_j, x_j\} \\ &= \sum_{j\in J}\sum_{i\in I_j} \mu_{i,j}\{T_{i,j}a_j, T_{i,j}b_j, T_{i,j}x_j\} \\ &= \{\sum_{j\in J}\sum_{i\in I_j} t_{i,j}a_j, \sum_{j\in J}\sum_{i\in I_j} T_{i,j}b_j, \sum_{j\in J}\sum_{i\in I_j} \mu_{i,j}T_{i,j}x_j\} \\ &= \{Ta, Tb, T_\mu x\}. \end{aligned}$$

Similarly,

$$T_\mu\{a,x,b\} = \{Ta, T_\mu x, Tb\}.$$

Thus $(T, T_\mu) : (U, U_p) \to (V, V_p(\mu))$ is a morphism in the category of modules. Let S_ν and $W_p(\nu)$ be constructed from the data $\{S_{k,j}; k \in K_j, j \in J\}, S, W$ and $\nu = \{\nu_{k,j}; k \in K_j, j \in J\}$. Then $(S, S_\nu) : (U, U_p) \to (W, W_p(\nu))$ is a morphism. Thus $(ST^{-1}, S_\nu T_\nu^{-1}) = (S, S_\nu) \circ (T, T_\mu)^{-1}$ is a morphism from $(V, V_p(\mu))$ onto $(W, W_p(\nu))$. □

Our next goal is to show that every contractive projection defined on a member of $\mathcal{C}_p$ is a conditional expectation. To explain this consider first a classical Cartan factor U, and let $T : U \to B(H)$ be an irreducible, faithful triple-representation. Let $V = T(U)$ and let E be the orthogonal projection from C_2 onto V_2. As explained in Section 2 (in the paragraph beginning after equation (2.24)), E extends to a contractive projection from $B(H)$ onto V and is also a contractive projection from C_p onto V_p, $1 \le p \le \infty$. The contractive projections E constructed in this way are examples of *conditional expectations.* (The terminology is justified by Theorem 8.17 and Remark 8.18 below).

In general, let $\{U_j\}_{j\in J}, U, \{T_{i,j}; i \in I_j, j \in J\}, T, V, \mu = \{\mu_{i,j}; i \in I_j, j \in J\}, T_\mu$ and $V_p(\mu)$ be as in the discussion preceeding Definition 8.4. Let $E_{i,j}$ be the conditional expectation onto $T_{i,j}(U_j)$. Define for $x \in C_p$

$$E_p(\mu)x = \sum_{j\in J}\sum_{i\in I_j} \mu_{i,j} T_{i,j}\Big(\sum_{k\in I_j} \mu_{k,j}^{p-1} T_{k,j}^{-1} E_{k,j} x\Big). \tag{8.10}$$

For $x \in B(H)$ define

$$E_\infty(\mu)x = \sum_{j\in J}\sum_{i\in I_j} T_{i,j}\Big(\sum_{k\in I_j} \mu_{k,j}^{p} T_{k,j}^{-1} E_{k,j} x\Big). \tag{8.11}$$

Theorem 8.17. **(a)** *$E_p(\mu)$ is the contractive projection from C_p onto $V_p(\mu)$;*

(b) *$E_\infty(\mu)$ is a contractive projection from $B(H)$ onto V;*

(c) *For every $a, b \in V$ and $x \in C_p$,*

$$E_p(\mu)\{a,b,x\} = \{a,b,E_p(\mu)x\}, \quad E_p(\mu)\{a,x,b\} = \{a,E_p(\mu)x,b\};$$

(d) *For every $a, b \in V$ and $x \in B(H)$,*

$$E_\infty(\mu)\{a,b,x\} = \{a,b,E_\infty(\mu)x\}, \quad E_\infty(\mu)\{a,x,b\} = \{a,E_\infty(\mu)x,b\}.$$

Proof: **(a)** Consider $T_{\mu^{p-1}}$ as an isometry of U_q into C_q $\left(\frac{1}{p}+\frac{1}{q}=1\right)$. Then the adjoint map $T^*_{\mu^{p-1}} : C_p \to U_p$ is given by

$$T^*_{\mu^{p-1}} x = \sum_{j\in J}\sum_{i\in I_j} \mu_{i,j}^{p-1} T_{i,j}^{-1} E_{i,j} x.$$

Thus

$$T^*_{\mu^{p-1}} T_\mu = I_{U_p}, \quad E_p(\mu) = T_\mu T^*_{\mu^{p-1}}.$$

This shows that $E_p(\mu)$ is the contractive projection from C_p onto $V_p(\mu)$.

(b) Similarly, considering T_{μ^p} as an isometry of U_1 into C_1, we see that

$$T^*_{\mu^p} T = I_U, \quad E_\infty(\mu) = TT^*_{\mu^p}.$$

Thus $E_\infty(\mu)$ is a contractive projection from $B(H)$ onto V.

(c) Let $y = \sum_{j \in J} y_i$, $z = \sum_{j \in J} z_j \in U$ and let $a = Ty, b = Tz$. It is easy to see by (2.9) that

$$E_{k,\ell}\{T_{i,j} y_j, T_{i,j} z_j, x\} = \delta_{j,\ell}\delta_{i,k}\{T_{i,j} y_j, T_{i,j} z_j, E_{i,j} x\}.$$

Thus, since $T_{i,j}$ are triple-isomorphisms,

$$\begin{aligned} E_p(\mu)\{a,b,x\} &= \sum_{j \in J}\sum_{k \in I_j} \mu_{k,j} T_{k,j}\left(\sum_{i \in I_j} \mu_{i,j}^{p-1} T_{i,j}^{-1}\{T_{i,j} y_j, T_{i,j} z_j, E_{i,j} x\}\right) \\ &= \sum_{j \in J}\sum_{k \in I_j} \mu_{k,j} T_{k,j}\{y_i, z_j, \sum_{i \in I_j} \mu_{i,j}^{p-1} T_{i,j}^{-1} E_{i,j} x\} \\ &= \{Ty, Tz, E_p(\mu)x\} = \{a, b, E_p(\mu)x\}. \end{aligned}$$

The proof of the formula $E_p(\mu)\{a,x,b\} = \{a, E_p(\mu)x, b\}$ as well as the proof of (d) is similar. □

Remark 8.18. Properties (c) and (d) in Theorem 8.17 justify the name *conditional expectations* for the projections $E_p(\mu)$ and $E_\infty(\mu)$. Also, (c) and (d) mean that $E_p(\mu)$ and $E_\infty(\mu)$ commute with the operators $D(a,b) = \{a,b,\cdot\}$ and $Q(a,b) = \{a,\cdot,b\}$ for all $a, b \in V$.

The next application is to *the structure of the powers of the members* of $\mathcal{C}_p$. If X is a subspace of C_p and $0 < r < \infty$, let $X^r = \{x^r; x \in X\}$ where $x^r := v(x)|x|^r$ as usual. In general X^r is a subset of $C_{p/r}$ which need not be a linear subspace. For the special value $r = p-1$, the Main Theorem says that $X \in \mathcal{C}_p$ if and only if $N_p(X) = X^{p-1} \in \mathcal{C}_q$ $\left(\frac{1}{p} + \frac{1}{q} = 1\right)$. We can strengthen this as follows.

Proposition 8.19. *Let $1 < p < \infty, p \neq 2, 0 < r < p, r \neq p/2$, and let X be a subspace of C_p. Then $X \in \mathcal{C}_p$ if and only if $X^r \in \mathcal{C}_{p/r}$.*

Proof: Suppose $X \in \mathcal{C}_p$, and write $X = V_p(\mu) = T_\mu(U_p)$. Then $X^r = T_{\mu^r}(U_{p/r})$ and so $X^r \in \mathcal{C}_{p/r}$. The other implication follows in the same way. □

9. Families of contractive projections and concluding remarks.

In this section we apply the Main Theorem to families of contractive projections. We include the cases $p = 1, \infty$ in this discussion. For $1 \leq p < \infty$, $p \neq 2$ the ranges of contractive projections in C_p have the same description. This follows from the Main Theorem for $1 < p < \infty, p \neq 2$, and from [AF2] for $p = 1$. As in Section 8 this class of subspaces is denoted by $\mathcal{C}_p$, but in this section $p = 1$ is also permitted. The description of ranges of contractive projections in C_∞ is a little more complicated and is described in full in [AF2]. This description will not be used here, as the results in case $p = \infty$ will follow from the case $p = 1$ by duality.

A contractive projection P on a Banach space X is *bicontractive* if $I - P$ is also contractive. An isometry T of X of period 2 (i.e. $T^2 = I$) is called a *symmetry*. Every symmetry T gives rise to a bicontractive projection $P = (I + T)/2$ with a complement $I - P = (I - T)/2$. If P is a bicontractive projection on X, then in general $T = 2P - I = P - (I - P)$ need not be a symmetry. However, in the context of JB^*-triples every bicontractive projection generates a symmetry this way [FR3]. C_p is also well behaved in this respect.

Theorem 9.1. *Let* $1 \leq p \leq \infty$, $p \neq 2$.

(a) *An isometry T of C_p is a symmetry if and only if it has one of the following forms. Either*

$$Tx = uxv; x \in C_p \tag{9.1}$$

with u, v symmetries of the underlying Hilbert space H; or

$$Tx = \pm wx^*w; \; x \in C_p \tag{9.2}$$

where w is a conjugate-linear isometry of H.

(b) *A contractive projection P on C_p is bicontractive if and only if it has one of the following forms. Either*

$$Px = pxq + (1-p)x(1-q); \; x \in C_p \tag{9.3}$$

with p, q projections on H which are not simultaneously trivial, or

$$Px = (x \pm wx^*w)/2; \; x \in C_p \tag{9.4}$$

where w is a conjugate-linear isometry of H.

(c) *P is a bicontractive projection on C_p if and only if $T := 2P - I$ is an isometric symmetry of C_p.*

Remark 9.2. Formula (9.2) is equivalent to $Tx = \pm ux^T\bar{u}$, where x^T is the transpose of x relative to a fixed orthonormal basis $\{e_j\}$ of H and $\bar{u}$ is the operator on H whose matrix with respect to this basis of H is the conjugate of the matrix of u. If J is the conjugation on H associated with $\{e_j\}$ (namely $J(\sum_j \xi_j e_j) = \sum_j \bar{\xi}_j e_j$) then $\bar{u} = JuJ$ and so $ux^T\bar{u} = wx^*w$ with $w = uJ$. Formula (9.2) is best understood in the framework of the theory of *Jordan pairs*, see [Lo].

Proof of Theorem 9.1. It is clear that if T is given by either (9.1) or (9.2), then T is a symmetry. If P is given by (9.3) then $P = (I+T)/2$ where T is given by (9.1) with $u = 2p-1, v = 2q-1$, Thus P is a bicontractive projection. Similarly, if P is given by (9.4) then $P = (I+T)/2$ where T is given by (9.2) and again P is a bicontractive projection. These arguments show also that (b) implies (c).

To prove the other implication in (a), we need the following result.

Theorem 9.3. [A1] *Every isometry T of $C_p, 1 \leq p \leq \infty, p \neq 2$, is of one of the following forms. Either*

$$Tx = uxv; \quad x \in C_p \tag{9.5}$$

or

$$Tx = ux^T v; \quad x \in C_p \tag{9.6}$$

where u, v are unitary operators and x^T is the transpose of x relative to some fixed orthonormal basis in H.

Let T be an isometric symmetry of C_p, By Theorem 9.3, T is of one of the forms (9.5) or (9.6). If T is given by (9.5) then $x = T^2 x = u^2 x v^2$ for every $x \in C_p$. This implies that $u^2 = \lambda I$ and $v^2 = \bar{\lambda} I$ for some unimodular scalar λ. Define $u_1 = \bar{\lambda}^{1/2} u$ and $v_1 = \lambda^{1/2} v$. Then $u_1^2 = v_1^2 = I$ and $Tx = u_1 x v_1$, $x \in C_p$. If T has the form (9.6), then $x = T^2 x = uv^T x u^T v$ for all $x \in C_p$. This implies that $v = \pm \bar{u}$, and so T is given by (9.2) with $w = uJ$, see Remark 9.2.

It remains to prove that every bicontractvie projection on C_p is of one of the forms (9.3) or (9.4). We need an auxiliary result which is of independent interest. It will be convenient to introduce some notation. If $Y \in C_p$ is indecomposable let q_1, q_2 be the smallest projection on H so that $q_1 Y q_2 = Y$ and define $P_2(Y)x = q_1 x q_2, P_1(Y)x = q_1 x(1 - q_2) + (1 - q_1) x q_2$ and $P_0(Y)x = (1 - q_1)x(1 - q_2)$. We call $P_j(Y), j = 0, 1, 2$, the *Peirce projections of* Y. They can be expressed in terms of the Peirce projections of the element of some cog-grid in Y.

Lemma 9.4. *Let U be a Cartan factor of rank r and let $Q : U_p \to U_p$ be a contractive projection* $(1 \leq p \leq \infty, p \neq 2)$. *If $Q \neq I$ then* $\dim(kerQ) \geq r - 1$.

Proof: The case $r \leq 2$ is trivial, so we assume $r \geq 3$. Thus U is of one of the types I, II or III. Assume first that $1 < p < \infty, p \neq 2$. If $Z := R(Q)$ is indecomposable and $P_0(Z)U_p = \{0\}$, we get by Theorem 2.4, that U cannot be of one of the types II, III. Hence $U = U(I_{r,m}) = B(H, H \oplus K)$ with H, K Hilbert space and $\dim H = r, \dim H \oplus K = m$. There are 3 possibilities for Z: either $Z = \{x \in B(H,H);\ x^T = x\}$ or $Z = \{x \in B(H,H);\ x^T = -x\}$ or $Z = B(H, H \oplus K_0)$ where K_0 is a proper subspace of K and x^T is the transpose of x relative to some orthonormal basis. If $r = \infty$ this shows that $\dim(kerQ) = \infty$. If $r < \infty$ then examination of the 3 cases shows that $\dim(kerQ)$ is estimated from below by $r(m - (r-1)/2)$, or $r(m-(r+1)/2)$ or r respectively. Thus $\dim(kerQ) \geq r$.

The other alternative is that Z contains an indecomposable summand, say Y, with $P_0(Y)U_p \neq \{0\}$. Clearly, $QP_1(Y) = 0$, hence $\dim(kerQ) \geq \dim(P_1(Y))$. If $r = \infty$ then $\dim P_1(Y) = \infty$. If $r < \infty$ then $\rho := rank(Y) \leq r-1$, and $\dim P_1(Y) \geq \rho(r-\rho) \geq r-1$.

The proof in case $p = 1$ is the same since Theorem 2.4 holds for $p = 1$ by the results of [AF2]. The proof in case $p = \infty$ follows from the case $p = 1$ by standard duality arguments: if Q is a contractive projection on U_∞ then Q^* is a contractive projection on $(U_\infty)^* = U_1$ and $\dim(kerQ) = \dim(kerQ^*)$. □

Completion of the Proof of Theorem 9.1

We assume that $1 \leq p < \infty, p \neq 2$. The proof in case $p = \infty$ follows from the case $p = 1$ by duality. Let P be a bicontractive projection on C_p with $X = R(P)$ and let x be an atom of X. Denote $U(x) := P_2(x)B(H)$. Then $U(x) = U(I_{r,r})$ for $r = rank(x)$, and by Theorem 4.1, $U_p(x) := P_2(x)C_p$ is invariant under both P and $I-P$. Let $Q = (I-P)_{|U_p(x)}$. Then $kerQ = \mathbf{C}x$. Thus Lemma 9.4 yields $1 = \dim(kerQ) \geq r-1$ and so $r \leq 2$.

Suppose that P has an atom x of rank 2, i.e., $rank(U(x)) = rank(U_p(x) = 2$. Since $\dim(PU_p(x)) = 1, \dim((I-P)U_p(x)) = 3$. Apply Theorem 2.4 to $U_p(X)$ and $I-P$ to conclude that $(I-P)U_p(x) = U_p(III_2)$. Let Y be the indecomposable component of X containing x and let $Z = P_2(Y)$. Then $Z = U_p(I_{n,n})$ for some $n \leq \infty$ and Z is invariant under P and $I-P$. Moreover, $Y = PZ = U_p(II_n)$ and $(I-P)Z = U_p(III_n)$. Clearly $P_1(Y) = 0$, hence $P_0(Y) = 0$ as well. Thus $Z = C_p$ and P is the canonical projection on $U_p(II_n)$ (i.e. $x \mapsto (x-x^T)/2$), possibly conjugated by an isometry of C_p. Using Theorem 9.3 we see that $P = (x - wx^*w)/2$ for some conjugate-linear isometry w of H.

Similarly, if $I - P$ has an atom x of rank 2, then the same proof yields $Px = (x + wx^*w)/2$ for some conjugate-linear isometry w.

The remaining case is that all the atoms of P and $I - P$ have rank 1. By the classification of ranges of contractive projections (Theorem 7.11) every indecomposable summand of either P or $I - P$ is a Cartan factor of type I. To complete the proof we need to show only that the number of indecomposable summands of P and $I - P$ is at most 2; this will yield formula (9.3). Indeed, if P has 3 or more indecomposable summands let x_1, x_2, x_3 be atoms of P from different indecomposable summands. Let $U = P_2(x_1 + x_2 + x_3)B(H)$. Then $U = U(I_{3,3})$, the space of 3×3 matrices, is invariant under both P and $I - P$. Thus, P restricted to $U(I_{3,3})$ is the projection onto the diagonal of the 3×3 matrices which annihilates the off-diagonal entries. Thus $(I - P)|_{U(I_{3,3})}$ is not contractive. This contradiction shows that the number of indecomposable summands of P and $I - P$ is at most 2, and concludes the proof of Theorem 9.1. □

The next result is an estimation of the gap between two members of the class $\mathcal{C}_p$ of ranges of contractive projections on C_p.

Lemma 9.5. *Let $1 \leq p < \infty, p \neq 2$. Let $X, Y \in \mathcal{C}_p$ satisfy the following conditions:*

(i) *X has an indecomposable summand X_1 with $rank(X_1) = r$;*

(ii) *$X \subset Y$ and X_1 is not a summand in Y.*

Then $\dim(Y/X) \geq r - 1$.

Proof: The lemma is trivially true for $r = 1, 2$. Thus we will assume $3 \leq r$. We use the full classification of indecomposable members of $\mathcal{C}_p$ (provided for $p = 1$ by [AF2], and for $1 < p < \infty$, $p \neq 2$, by Theorem 2.4). Pick an atom x of X_1. If x is not an atom of Y, write $x = y_1 + y_2$ with $y_1, y_2 \in Y$ non-zero orthogonal elements.

It follows that

$$\dim(Y/X) \geq \dim[(P_2(X_1)Y)/X] \geq \dim(X_1) \geq r.$$

If x is an atom of Y then it generates an indecomposable summand Y_1 which contains X_1. Clearly, X_1 is triple-isometric to U_p where U is a Cartan factor of one of the types $I_{n,m}$ (with $\min(n,m) = r$), II_r or III_r. Y_1 is triple-isometric to V_p, where V is a Cartan factor of the same type as U with $U \subset V$, $U \neq V$. It follows by examining each of the three cases that

$$\dim(Y/X) \geq \dim(Y_1/X) = \dim(V_p/U_p) \geq r - 1. \qquad \square$$

Remark 9.6. Lemma 9.5 holds for $p = \infty$ with some necessary modifications. X has the structure of a JB^*-triple, but it need not be a subtriple of $B(H)$. So the rank r is the abstract rank of X_1 coming from the JB^*-triple structure, and not from the JC^*-triple structure of $B(H)$. Alternatively, if P is a contractive projection on C_∞ with range X_1, define r to be the rank of $R(P^*)$ ($\subset C_1$). Similarly, interpret $P_1(X_1)$ as $P_1(R(P^*))$. Now the statements of Lemma 9.5 are correct in C_∞, and the obvious proof is by duality.

We apply now Lemma 9.5 to monotone Schauder decompositions in C_p. Recall that a sequence of subspaces $\{Y_n\}$ of a Banach space X is a *Schauder decomposition* of X if every element $x \in X$ has a unique expansion $x = \sum_n x_n$ with $x_n \in Y_n$ for every n. In this case the projections P_n on $X_n := \sum_{k=1}^n Y_k$ defined by $P_n(\sum_k x_k) = \sum_{k=1}^n x_k$, $n = 1, 2, \cdots$, are necessarily bounded. Moreover $b := \sup_n \|P_n\| < \infty$. If the constant b equals 1 the decomposition $\{Y_n\}$ is said to be *monotone*. This is the case precisely if all the projections P_n are contractive. The *gaps* in the Schauder decomposition $\{Y_n\}$ are $\dim(Y_{n+1}) = \dim(X_{n+1}/X_n), n = 0, 1, 2, \ldots$ (where $X_0 = \{0\}$).

THEOREM 9.7. *Let U be a Cartan factor of rank r. Then any monotone*

Schauder decomposition $\{Y_n\}$ *of* U_p $(1 \leq p \leq \infty, p \neq 2)$ *satisfies*

$$\sup_n[\dim Y_n] \geq r - 1. \tag{9.7}$$

Proof: Let $X_n := \sum_{k=1}^{n} Y_k, n = 0, 1, 2, \ldots$. We have to show that $\sup_n[\dim(X_{n+1}/X_n)] \geq r-1$. We assume that $1 \leq p < \infty, p \neq 2$. The case $p = \infty$ follows from the case $p = 1$ by duality. As (9.7) is trivial for $r \leq 2$, we assume $3 \leq r$. Hence, either $U = U(I_{r,m})$ for $r \leq m$, or $U = U(II_r)$ or $U = U(III_r)$. If $r < \infty$ and either $U = U(II_r)$ or $U = U(III_r)$ then U is finite dimensional and hence the sequence $\{X_n\}$ is finite, say $X_N = U_p$. It follows by Lemma 9.4 that $\dim(X_N/X_{N-1}) = \dim(ker P_{N-1}) \geq r - 1$. The case $U = U(I_{r,m}), r \leq m$ and $m < \infty$ is handled similarly.

In all the remaining cases $\dim U_p = \infty$, and we can assume that $\{X_n\}$ is infinite and that $\dim X_n < \infty$ for every n. If $U = U(I_{r,\infty})$, $r < \infty$, there exists an index n so that $X_n = U(I_{r,k})$ for $r \leq k < \infty$. Indeed, choose n_1 so that X_{n_1} contains an element x of rank r. Then $\{P_2(x)P_j\}_{j=n_1}^{\infty}$ is a non-decreasing sequence of contractive projections in the finite dimensional space $U(I_{r,r}) = P_2(x)C_p$. Let n be the first number so that $P_2(x)P_nC_p = P_2(x)X_n = U(I_{r,r})$. By Theorem 2.4 (in case $1 < p < \infty, p \neq 2$) or [AF2] for $p = 1$, we get that $X_n = U(I_{r,k})$ for some $r \leq k < \infty$. The same reasoning gives $X_{n+1} = U(I_{r_1 k_1})$ with $k_1 > k$, hence $\dim(X_{n+1}/X_n) = r(k_1 - k) \geq r$.

It remains to consider the case where $r = \infty$. We have to show that $\{X_n\}$ has gaps of arbitrary large dimensions. Fix a positive integer ℓ. There exists a number n so that X_n has an indecomposable summand, say Z_n, with $\text{rank}(Z_n) \geq \ell$. Let $m > n$ be the first integer so that Z_n is not a summand in X_m. Notice that Z_n is a summand of X_{m-1}. We apply Lemma 9.5 with X_{m-1}, Z_n and X_m in the places of X, X_1 and Y respectively, and obtain $\dim(X_m/X_{m-1}) \geq \ell$. □

Recall that a *Schauder basis* for a Banach space X is a sequence $\{x_n\}_{n=1}^{\infty}$ so that every $x \in X$ has a unique expansion of the form $x = \sum_{n=1}^{\infty} \lambda_n x_n$ with $\lambda_n \in \mathbf{C}$. In this case, one defines $Y_n = \mathbf{C}x_n$, and $\{Y_n\}$ is a Schauder decomposition of X. The basis $\{x_n\}_{n=1}^{\infty}$ is said to be *monotone* if $\{Y_n\}$ is monotone.

Corollary 9.8 *Let U be a clasical Cartan factor and let $1 \leq p \leq \infty, p \neq 2$. Then U_p has a monotone basis if and only if $rank(U) \leq 2$.*

Proof: Let $r = rank(U)$. If $r = 1$ then $U = U(I_{1,n})$ for some $1 \leq n \leq \infty$ and U_p is isometric to a Hilbert space. Thus every orthonormal basis of U_p is a monotone basis.

If $r = 2$, then $U = U(IV_n)$ for $3 \leq n \leq \infty$. If $n = \infty$ then $U_p = \{0\}$; this follows easily from the description of the representations of $U(IV_n)$ in $B(H)$ given in [AF2]. If $n < \infty$, then U_p has a monotone basis consisting of the elements of a cog-grid which corresponds to the increasing sequence of Cartan factors $\{U(IV_k)\}_{k=1}^{n}$.

If $r \geq 3$ and $\{Y_n\}$ is any monotone Schauder decomposition of U_p, then by Theorem 9.7, $\dim(Y_n) \geq 2$ for at least one number n. In particular, U_p does not have a monotone basis. □

Next let $\{U_j\}_{j\in J}$ be an orthogonal family of Cartan factors in $B(H)$ and let

$$U = \left(\sum_{j\in J} U_j\right)_{\infty} = \{x = \sum_{j\in J} x_j;\ x_j \in U_j,\ \sup_j \|x_j\| < \infty\}.$$

We would like to analyze the structure of a contractive projection P on $U_p = (\sum_{j\in J}(U_j)_p)_p$ with range $R(P) = X$, $1 \leq p < \infty, p \neq 2$. If $E_j : B(H) \to U_j$ is the conditional expectation and $E = \sum_{j\in J} E_j$, then E is a contractive projection from $C_p = C_p(H)$ onto U_p, thus PE is a contractive projection on C_p having the same range as P. Let $\{X_k\}_{k\in K}$ be the indecomposable summands of X.

For every $k \in K$ fix an atom x_k of X_k and write $x_k = \sum_{j\in J_k} x_{k,j}$ with $x_{k,j} = E_j x_k \in (U_j)_p$. Fix a $j \in J$ for which $x_{k,j} \neq 0$. Then $y_{k,j} := x_{k,j}/\|x_{k,j}\|_p$ is an atom of an indecomposable subspace $X_{k,j}$ of $(U_j)_p$ from the class $\mathcal{C}_p$. This follows from the detailed description of the class $\mathcal{C}_p$ in Theorem 2.4 for $1 < p < \infty, p \neq 2$, and in [AF2] for $p = 1$. Let $P_{k,j}$ be the canonical contractive projection from C_p

onto $X_{k,j}$ (for $1 < p < \infty, p \neq 2$, $P_{k,j}$ is unique, whereas for $p = 1$, $P_{k,j}$ is the unique contractive projection onto $X_{k,j}$ satisfying $P_{k,j}x = 0$ for every x which is orthogonal to all of $X_{k,j}$). Notice that the map $x \to E_j x/\|x_{k,j}\|$ is a triple-isometry of X_k onto $X_{k,j}$. If $x_{k,j} = 0$, define $X_{k,j} = \{0\}$ and $P_{k,j} = 0$. For each $j \in J$ define $X^{(j)} = \sum_{k\in K} X_{k,j}$ and $P^{(j)} = \sum_{k\in K} P_{k,j}|_{(U_j)_p}$. Then $P^{(j)}$ is a contractive projection on $(U_j)_p$ with $R(P^{(j)}) = X^{(j)}$.

Let Q be an another contractive projection on U_p with $Y := R(Q) \supset X$ and $Y \neq X$. Construct in an analogous way subspaces $Y^{(j)}$ of $(U_j)_p$ and contractive projections $Q^{(j)}$ from $(U_j)_p$ onto $Y^{(j)}$.

Lemma 9.9. *In the situations just described, for every $j \in J$, $Y^{(j)} \supset X^{(j)}$ and*

$$\dim(Y^{(j)}/X^{(j)}) \leq \dim(Y/X). \tag{9.8}$$

Proof: Let $\{Y_\ell\}_{\ell\in L}$ be the indecomposable summands of Y. For every $k \in K, X_k \subset \sum_{\ell\in L_k} Y_\ell$ where the sets $\{L_k\}_{k\in K}$ are disjoint. Let also $\tilde{L} = L \setminus \cup_{k\in K} L_k$. For every $j \in J, \ell \in L$ let $Y_{\ell,j} = E_j(Y_\ell)$. Then $Y^{(j)} = \sum_{\ell\in L} Y_{\ell,j}$ and $X_{k,j} \subset \sum_{\ell\in L_k} Y_{\ell,j}$. Thus

$$X^{(j)} = \sum_{k\in K} X_{k,j} \subset \sum_{k\in K}\sum_{\ell\in L_k} Y_{\ell,j} \subset \sum_{\ell\in L} Y)_{\ell,j} = Y^{(j)},$$

and

$$Y^{(j)}/X^{(j)} \equiv \sum_{\ell\in\tilde{L}} Y_{\ell,j} + \sum_{k\in K}(\sum_{\ell\in L_k} Y_{\ell,j})/X_{k,j}.$$

It follows that

$$\begin{aligned}\dim(Y^{(j)}/X^{(j)}) &= \sum_{\ell\in\tilde{L}} \dim(Y_{\ell,j}) + \sum_{k\in K} \dim(\sum_{\ell\in L_k} Y_{\ell,j}/X_{k,j}) \\ &\leq \sum_{\ell\in\tilde{L}} \dim(Y_\ell) + \sum_{k\in K} \dim(\sum_{\ell\in L_k} Y_\ell/X_k) = \dim Y/X. \square\end{aligned}$$

Theorem 9.10. *Let $\{U_j\}_{j\in J}$ be an orthogonal family of Cartan factors in $B(H)$, let $U = (\sum_{j\in J} U_j)_\infty$ and let $1 \leq p < \infty, p \neq 2$.*

(a) *Let $\{Y_n\}$ be a monotone Schauder decomposition of U_p. Then*

$$\sup_n\{\dim Y_n\} \geq \sup_{j\in J}\{rank(U_j) - 1\}. \tag{9.9}$$

(b) *U_p has a monotone basis if and only if $rank(U_j) \leq 2$ for every j.*

Proof: As in Theorem 9.7, we consider only $1 \leq p < \infty, p \neq 2$. The case $p = \infty$ follows from the case $p = 1$ by duality. Let $X_n := \sum_{k=1}^{n} Y_k$, $n = 0, 1, 2\ldots.$ Combining Lemma 9.9 with Theorem 9.7, we get

$$\begin{aligned}\sum_n \dim(Y_n) &= \sup_n\{\dim(X_{n+1}/X_n)\} \\ &\geq \sup_n \sup_{j\in J}\{\dim(X_{n+1}^{(j)}/X_n^j)\} \\ &\geq \sup_{j\in J}\{rank(U_j) - 1\}.\end{aligned}$$

(b) If U_p has a monotone basis then (a) yields $rank(U_j) \leq 2$ for every j. If $rank(U_j) \leq 2$, then by Corollary 9.8 $(U_j)_p$ has a monotone basis. Thus U_p has a monotone basis as the ℓ_p-sum of spaces with monotone bases. □

CONCLUDING REMARKS AND OPEN PROBLEMS.

1. **Unconditionality of monotone Schauder decompositions of C_p.** Recall that a Schauder decomposition $\{Y_n\}_{n=1}^{\infty}$ of a Banach space X is *unconditional* if for every $x \in X$ with expansion $x = \sum_{n=1}^{\infty} x_n, x_n \in Y_n$, and every choice of signs $\epsilon_n = \pm 1$, the series $\sum_{n=1}^{\infty} \epsilon_n x_n$ converges as well.

Problem 1. *Is every monotone Schauder decomposition of C_p $(1 < p < \infty, p \neq 2)$ unconditional?*

We conjecture that the answer is affiramative. The conjecture is supported by the very detailed information we obtained on the structure of contractive projections in C_p. The unconditionality of the *shell-decomposition*, corresponding to the increasing sequence of projections $P_n x = p_n x p_n$, where $\{p_n\}_{n=1}^{\infty}$ are increasing orthogonal projections on H with $p_n \to I$ strongly, is proved in [KP]. The unconditionality of monotone Schauder decompositions in $L_p(\mu)$-spaces is proved in [DO] and [PR].

2. **Extension of our results to other non-commutative L_p-spaces.**

Our techniques can be extended to study contractive projections in other non-commutative L_p-spaces. First, our results hold, with the same proofs, in the context of the C_p-ideals of the exceptional Cartan factors $U(V)$ and $U(VI)$ of dimensions 16 and 27 respectively. Therefore our Main Theorem and its corollaries hold in the context of the C_p spaces associated with general atomic JBW^*-triples. The notation used in Section 8 which avoids the tensor-product notation (which is meaningless in general JB^*-triples) is particularly convenient in this generalization. Another case where our technique can be applied is that of the $L_p(M, \tau)$ spaces, where M is a (von-Neumann) factor of type II_1 or II_∞ and τ is the canonical trace on M. The results of Section 3 and 4 hold also in this case with similar proofs. However, in Sections 5, 6 and 7 our proofs use in an essential way the atomic nature of C_p (i.e., the fact that the underlying von-Neumann algebra, $B(H)$, is a type I factor). Thus new ideas are needed in order to extend our Main Theorem to this context.

3. **Connection between ranges of contractive projections in C_p and functional calculus of homogeneous functions of degree 1.**

One of the characterizations of the spaces in $\mathcal{C}_p$ $(1 < p < \infty, p \neq 2)$, i.e. ranges of contractive projections in C_p, is the N-convexity condition. Namely, a subspace X of C_p belongs to $\mathcal{C}_p$ if and only if $(x^{p-1} + y^{p-1})^{1/(p-1)} \in X$. Here $x^\alpha = v(x)|x|^\alpha$ as usual.

Problem 2. *Let X be a subspace of C_p and let $0 < r < \infty, r \neq 1$. Suppose* *that* $(x^r + y^r)^{1/r} \in X$ *for every* $x, y \in X$. *Must* $X \in \mathcal{C}_p$*?*

If $X \in \mathcal{C}_p$ has the form $X = V_p(\mu) = T_\mu(U_p)$, then for $x = T_\mu a, y = T_\mu b \in X$ one has $(x^r + y^r)^{1/r} = T_\mu(a^r + b^r)^{1/r} \in X$, because $(a^r + b^r)^{1/r} \in U_p$. We conjecture that the answer to Problem 2 is affirmative. The conjecture is supported by the results of Sections 3 and 4. One can consider other homogeneous functions of two variables in Problem 2 (for instance, operator means) and study the corresponding problem. We expect an affirmative answer also in this generality.

4. **Connection between ranges of contractive projections in C_p and homogeneous Schur multipliers of degree 0.**

Let $x \in C_p$ have a Schmidt series $x = \sum_j a_j v_j$. Let $\{P_{i,j}(x)\}_{0 \leq i \leq j}$ be the joint Peirce projections associated with $\{v_j\}$, and consider Schur multipliers of the form $M_x = \sum_{0 \leq i \leq j} m_{i,j}(\alpha) P_{i,j}(x)$ where the $m_{i,j}(\alpha)$ functions of the $\{\alpha_j\}$ which are homogeneous of degree 0. One of our major tools, developed in Section 3 and 4 is the fact that if $X \in \mathcal{C}_p$ and $x \in X$ then any Schur multiplier M_x of the above form, which is bounded on C_p, maps X into itself. In addition, X is closed under the "local conjugation" operator $Q(v(x))y = \{v(x), y, v(x)\}$. We conjecture that these two properties - closure under homogeneous "local" Schur multipliers of degree 0 and closure under the local conjugations is sufficient for X to be in $\mathcal{C}_p$. We Remark that the homogeneous local Schur multipliers of degree 0 appear in the context of Tomita-Takesaki theory on Jordan algebras [HH] as the modular operators. These operators can be interpreted as the local Schur multiplier corresponding to the ratio between the geometric and the arithmetic means, and it is thus homogeneous of degree 0.

5. **Connection between contractive projections on C_p and symmetric domains.**

Let D be a domain in a Banach space Z and assume that $0 \in D$ and that D is circular. D is called *symmetric* if the group $Aut(D)$ of all biholomorphic automorphisms of D acts transitively on D, i.e. D is *homogeneous* with respect to $Aut(D)$. By a fundamental Theorem of W. Kaup ([K1], see also [U1], [U2] and [Lo]), a complex Banach space Z is a JB^*-triple if and only if its open unit ball $D(Z)$ is a bounded symmetric domain. Let U be a classical Cartan factor and let $D(U)$ be its open unit ball. For $1 \leq p < \infty$ let $D_p(U) = D(U) \cap C_p$; $D_p(U)$ is an unbounded domain in U_p. Let $G = Aut(D(U))$ and let $G_p = \{g \in G; g(0) \in C_p\}$. It is easy to see that the members of G_p map $D_p(U)$ biholomorphically onto itself and that G_p acts transitively on $D_p(U)$. Thus $D_p(U)$ *is a symmetric domain.* Using the Main Theorem and the results of [AF2] we conclude that *if* $X \in \mathcal{C}_p$ $(1 \leq p \leq \infty)$ *is indecomposable then its open unit ball* $D(X)$ *is a symmetric domain.*

Problem 3. *Let* X *be an indecomposable subspace of* C_p $(1 \leq p \leq \infty, p \neq 2)$ *of rank* ≥ 2*, whose open unit ball is a symmetric domain. Must* $X \in \mathcal{C}_p$*?*

We conjecture that the answer is affirmative.

REFERENCES

[A1] J. Arazy, Isometries of C_p, Israel J. Math., Vol. 22 (1977), 247-256.

[A2] J. Arazy, Certain Schur-Hadamard multipliers in the spaces C_p, Proc. AMS 86(1982), 59-64.

[A3] J. Arazy, Some remarks on interpolation theorems and the boundness of the triangular projection in unitary matrix spaces, Integral Equations and Operator Theory, 1 (1978), 453-495.

[ABF] J. Arazy, T. Barton and Y. Friedman, Operator differentiable functions, Integral Equations and Operator Theory 13 (1990), 461-487.

[AF1] J. Arazy and Y. Friedman, Isometries of $C_p^{n,m}$ into C_p, Israel J. Math. 26 (1977), 151-165.

[AF2] J. Arazy and Y. Friedman, Contraction projections in C_1 and C_∞, Memoirs AMS No. 200, (1978).

[AF3] J. Arazy and Y. Friedman, Contractive projections in C_p (research announcement) to appear in C. R. Acad. Sci. Paris.

[BF] R. Bonic and J. Frampton, Smooth functions on Banach manifolds, J. Math. Mech., Vol. 15 (1966), 877-898.

[BG] E. Berkson and T.A. Gillespie, Stechkin's theorem, transference, and spectral decompositions, J. Funct. Anal. 70 (1987), 140-170.

[BSI] M.S. Birman and M.Z. Solomyak, *Stieltjes double-integral operators, in* "Topics in Mathematical Physics," vol. 1, Consultants Bureau, New York, (1967), 25-54.

[BSII] M.S. Birman and M.Z. Solomyak, *Stieltjes double-integral operators. II, in* "Topics in Mathematical Physics," vol. 2, Consultants Bureau, New York, (1968), 19-46.

[BSIII] M.S. Birman and M.Z. Solomyak, *Stieltjes double-integral operators. III (Passage to the limit under the integral sign)*, (Russian) Prob. Mat. Fiz. No. 6 (1973), 27-53.

[Da] M. Day, Normed linear spaces, 3rd edition, Springer Verlag (1973).

[DF] T. Dang and Y. Friedman, Classification of JBW^*-triple factors and applications, Math. Scand. 61 (1987), 292-330.

[Di] J. Diestel, Geometry of Banach spaces: selected topics, Lecture Notes in Math., Springer-Verlag (1975), No. 485.

[DK] J.L. Daleckii and S.G. Krein, Integration and differentiation of functions of hermitian operators and applications to the theory of perturbations, A.M.S.

Translations (2) 47 (1965), 1-30.

[Do] H.R. Dowson, Spectral theory of linear operators, Academic Press (1978).

[DO] L.E. Dor and E, Odell, Monotone bases in L_p, Pacific Journal of Mathematics, 60(1975), 51-61.

[DS1] N. Dunford and J.T. Schwartz, Linear Operators, Vol. 1: General theory, Pure and Applied Mathematics, Interscience Publishers, New York (1958).

[DS2] N. Dunford and J.T. Schwartz, Linear Operators, Vol. 2: Spectral Theory, Pure and Applied Mathematics, Interscience Publishers, New York (1963).

[ES] E. Effros and E. Stromer, Positive projections and Jordan structure in operator algebras, Math. Scand. 45 (1979), 127-138.

[F1] Y.B. Farforovskaya, Examples of a Lipschitz function of self-adjoint operators that gives a nonnuclear increment under a nuclear perturbation, J. Soviet. Math. 4 (1975), 426-433.

[F2]Y.B. Farforovskaya, An estimate of the norm $|f(B) - f(A)|$ for self-adjoint operators A and B, J. Soviet Math. 14 (1980), 1133-1149.

[F3] Y.B. Farforovskaya, An estimate of the difference $f(B)-f(A)$ in the classes γ_p, J. Soviet Math. 8 (1977), 146-148.

[FR1] Y. Friedman and B.Russo, Solution of the contractive projection problem, J. Funct. Anal., 60 (1985), 56-79.

[FR2] Y. Friedman and B. Russo, Structure of the predual of a JBW^*-triple, J. Reine Agrew. Math. 356 (1985), 67-89.

[FR3] Y. Friedman and B.Russo, Conditional expectation and bicontractive projections on Jordan C^*-algebras and teir generalizations, Math. Z 194 (1987), 227-236.

[GK1] I.C. Gohberg and M.G. Krein, Introduction to the theory of linear nonselfadjoint operators, Translations of the AMS, Vol. 18, Providence, R.I., 1969.

[GK2] I.C. Gohberg and M.G. Krein, Theory and Applications of Volterra operators in Hilbert spaces, Translations of the AMS, Vol. 29, 1970.

[H] L.A. Harris, Generalization of C^*-algebras, Proc. London Math. Soc. (3) (1981), 331-361.

[HH] U. Haagerup and H. Hanche-Olsen, Tomita-Takesaki theory for Jordan algebras, J. Operator theory, 11 (1984), 343-364.

[Ho] G. Horn, Classification of JB^*-triples of type I, Math. Z., 196 (1987),271-291.

[K1] W. Kaup, A Rieman mapping theorem for bounded symmetric domains in complex Banach spaces, Math. Z. 183 (1983), 503-529.

[K2] W. Kaup, Contractive propjections on Jordan C^*-algebras and generalizations, Math. Scand. 54 (1984), 95-100.

[KP] S. Kwapien and A. Pelczynski, The main triangular projection in matrix spaces and its applications, Studia Math. 34 (1970), 433-68.

[L] H. E. Lacey, Isometric theory of classical Banach spaces, Springer Verlag, New York, (1974).

[Lo] O. Loos, Bounded symmetric domains and Jordan pairs, University of California, Irvine, 1977.

[M] Ch. McCarthy, C_p, Israel J. Math. 5 (1967), 249-271.

[Mc] K. McCrimmon, Compatible Peirce decompositions of Jordan triple system, Pacific J. Math. 103 (1982), 57-102.

[Mi] V.D. Milman, A few observations on the connection between local theory and some other fields, in Geometrical Aspects of Functional Analysis, Lecture notes in Mathematics, 1317, Springer-Verlag (1989), 283-289.

[MM] K. McCrimmon and K. Meyberg, Coordinatization of Jordan triple systems, Comm. in Algebra, 9 (14), (1981), 1495-1542.

[N] E. Neher, Jordan Triple Systems by the Grid Approach, Lecture Notes in Mathematics, Vol. 1280, Springer-Verlag (1987).

[P] V.V. Peller, Hankel operators in the perturbations theory of unitary and self-adjoint operators, Funct. Anal. Appl. 19 (1985), 111-123.

[Pa] B.S. Pavlov, On multidimensional integral operators, in Problems in Mathematical Physics Vol. 2: Linear Operators and Operator Equations (Ed. V.I. Smirnov), Consultant Bureau, New York (1971), 81-97.

[PR] A.Pelczinski and H.P. Rosenthal, Localization techniques in L_p-spaces, Studia Math. 52(1975), 263- 289.

[Si] B. Simon, Trace ideals and their applications, Cambridge University Press, 1979.

[So] B. Solel, Contractive projections onto bimodules of von Neumann algebras, preprint (1990).

[SS] M.Z. Solomyak and V.V. Stenkin, On one class of Stieltjes multiple-integral operators, in Problems i Mathematical Analysis Vol. 2: Linear Operators and Operator Equations (Ed. V.I. Smirnov), Consultant Bureau, New York (1971), 99-108.

[T1] N. Tomczak-Jaegerman, The moduli of smoothness and convexity and Rademacher averages of trace classes S_p $(1 \leq p < \infty)$, Studia Math. 50 (1974), 163-182.

[T2] N. Tomczak-Jaegerman, On the differentiability of the norm in trace classes S_p, Seminaire Maurey-Schwartz, Espaces L_p, Applications radinifiantes et geometrie des espaces de Banach, Exp. XXII, 9pp., Ecole polytechnique, Paris, 1974-5.

[U1] H. Upmeier, Symmetric Banach manifolds and Jordan C^*-algebras, North Holland Math. Studies, Vol. 104 (1985);

[U2] H. Upmeier, Jordan algebras in analysis, operator theory and quantum mechanics, CBMS-NSF Regional Conference series in Math. Amer. Math. Soc., Providence R.I., No. 67, 1987.

[Y] F. Yeadon, Isometries of non-commutative L^p-spaces, Math. Proc. Cambridge Phil. Soc. 90 (1981), 41-50.

Jonathan Arazy
University of Haifa
Haifa 31999, Israel
and
University of Kansas
Lawrence, Kansas 66045

Yaakov Friedman
Jerusalem College of Technology
P.O. Box 16031,Jerusalem 91160, Israel

MEMOIRS of the American Mathematical Society

General instructions to authors for

PREPARING REPRODUCTION COPY FOR MEMOIRS

> **For more detailed instructions send for AMS booklet, "A Guide for Authors of Memoirs." Write to Editorial Offices, American Mathematical Society, P.O. Box 6248, Providence, R.I. 02940-6248.**

MEMOIRS are printed by photo-offset from camera copy fully prepared by the author. This means that the finished book will look exactly like the copy submitted. Thus the author will want to use a good quality typewriter with a new, medium-inked black ribbon, and submit clean copy on the appropriate model paper.

Model Paper, provided at no cost by the AMS, is paper marked with blue lines that confine the copy to the appropriate size.

Special Characters may be filled in carefully freehand, using dense black ink, or **INSTANT** ("rub-on") **LETTERING** may be used. These may be available at a local art supply store.

Diagrams may be drawn in black ink either directly on the model sheet, or on a separate sheet and pasted with rubber cement into spaces left for them in the text. Ballpoint pen is not acceptable.

Page Headings (Running Heads) should be centered, in CAPITAL LETTERS (preferably), at the top of the page — just above the blue line and touching it.

LEFT-hand, EVEN-numbered pages should be headed with the AUTHOR'S NAME;

RIGHT-hand, ODD-numbered pages should be headed with the TITLE of the paper (in shortened form if necessary).

Exceptions: PAGE 1 and any other page that carries a display title require NO RUNNING HEADS.

Page Numbers should be at the top of the page, on the same line with the running heads.

LEFT-hand, EVEN numbers — flush with left margin;

RIGHT-hand, ODD numbers — flush with right margin.

Exceptions: PAGE 1 and any other page that carries a display title should have page number, centered below the text, on blue line provided.

FRONT MATTER PAGES should be numbered with Roman numerals (lower case), positioned below text in same manner as described above.

MEMOIRS FORMAT

> **It is suggested that the material be arranged in pages as indicated below.**
> **Note: Starred items (*) are requirements of publication.**

Front Matter (first pages in book, preceding main body of text).

Page i — *Title, *Author's name.

Page iii — Table of contents.

Page iv — *Abstract (at least 1 sentence and at most 300 words).

Key words and phrases, if desired. (A list which covers the content of the paper adequately enough to be useful for an information retrieval system.)

**1991 Mathematics Subject Classification*. This classification represents the primary and secondary subjects of the paper, and the scheme can be found in Annual Subject Indexes of MATHEMATICAL REVIEWS beginnning in 1990.

Page 1 — Preface, introduction, or any other matter not belonging in body of text.

Footnotes: *Received by the editor date.
Support information — grants, credits, etc.

First Page Following Introduction – Chapter Title (dropped 1 inch from top line, and centered). Beginning of Text.

Last Page (at bottom) – Author's affiliation.